Simplified
Mathematics Series 1

Simplified Mathematics Series 1

FOR SECONDARY SCHOOLS

{with answers}

OKOKO CHIDOZIE CHRISTIAN

N.Dip{Computer Sciences} B.Sc{Mathematics} PGD {Education}
Senior Mathematics Teacher and formerly Head of Department,
College of Islamic Sciences Gusau Zamfara State, Nigeria.

AND

DAVIDSON CHIDIEBERE OKOKO

MIFIA,MAAPG,MSEG,MSPE
M.Sc{ Geology}University of Ibadan,B.Sc{ Geology}
University of Calabar, Cross River State ,Nigeria.

To order additional copies of this book, contact:
Xlibris Corporation
1-888-795-4274
www.Xlibris.com
Orders@Xlibris.com
84583

Contents

DEDICATION

This work is dedicated to the one ETERNAL God, the author, finisher of knowledge and wealth.
It is heartily dedicated to my lovely wife, Joy.
Also it is dedicated to our parents: Mr. Aaron Okenna and Mrs. Comfort. Agbachionne Okoko { ***Both of the Blessed Memory***} my siblings: Harrison, Chinonyerem, Ezinne, Onyii , and Favour who ministered to me in substance and person when I have no one there except Father, the Most High.

Acknowlegement

This book owes too many people who have helped and encouraged me that **"Success is all about creating value"**

I am grateful to my Principal Alhaji Mukhtar Idris , students and staff of College of Islamic Sciences ,Gusau Zamfara State. Such staff include: Mr.Emeka Onyebueke {Dept. of Business Studies}, Ojeke Mathew{Dept.of English / Linguistics} Mal.Badamasi Muhammed {Dept. of Biological Sciences} etc.

My thanks goes to Alhaji Salihu Ibrahim, Mal Hadi, Alfa and other staff of New weekly Legacy Newspapers of Zamfara State. I am grateful to my friends: Chukwuemeka Ugochukwu Okorienta {Canteen Road, Motor Spare Part Dealer}Gusau, Mr. Ejike. Bro Basil, Ndubuisi Electricals ,Joseph Increase of Ahmadu Bello Way ,Isaiah Ogbu , Bro Gideon Felix and other members of Evangelical Fellowship in the Anglican Communion{EFAC}, Cathedral Church of Christ Gusau Zamfara State for some of their materials bought which inspired me to write this book.

Also thanks to my Priests: Rev.Chris. Ogbodo, Rev Obinna Nnebe, Rev. Bako Timothy, Rev.N.C Daniel, Evangelist John.O.John ,Evangelist Dr.E.O Titus and all Cathedral Choir Members who prayed tirelessly for my success, may the good God always be with you all ,Amen.

I will not forget showing my appreciation to **Mr.Joseph Onuh { Manager }C.O Onuh Business and Photocopying Centre ,Gusau, Mr.Peter Onuh and the entire staff of the above named Company.** Also to HONU Business and Photocopying Center , Opposite Sambo Secondary School Gusau,Zamfara State.

The immense work done by Miss Victoria Thomas and the staff of Al-Ilmu Business Centre requires a great commendations.

Great thanks goes to my immediate elder brother and co-author of my books *"Simplified Mathematics Series 1 for Secondary Schools with Answers"* *"Increase Your Financial Intelligence"* and *"Money Never Sleeps"*, Mr. Okoko Davidson Chidiebere of AAA Capital Market and Real Estate Corporation registered with the State of New York ,{U.S.A} for taking the pains to enlighten me and some of their material used.

In my task , I have been greatly assisted by a first rate team of academics who have worked alongside me in the development of this edition. Among whom are: Dr. Mrs. Ohakwe,{ Head of Department, Mathematics/Education F.C.E.T Gusau, Zamfara State Nigeria }, Alhaji Augie Hassan {Ph.D in view}, Department of Mathematics, Federal College of Education Technical{F.C.E.T} Gusau ,Zamfara State, Nigeria. To them I should express my sincere gratitude for all the detailed care and considerations they have rendered.

Preface

"Simplified Mathematics Series 1 for Secondary Schools with Answers" is designed to test among other things:

In view of the poor performances of candidates in the Senior Secondary Certificate Examination {SSCE}, National Examination Council {NECO}, General Certificate of Examination {GCE},O'Level and the Unified Tertiary Matriculation Examination {UTME} . . . It is clear that the objectives of this Series are not being realized to a satisfactory extent.

The aim of the author therefore is to provide a compilation of solutions to the status quo examinations bodies in line with the Mathematics curriculum and SSCE Syllabus **2011**. The book uses a conceptual approach in dealing with the following main themes:

- ✓ Circle.

- ✓ Bearing.

- ✓ Commercial /Financial Arithmetic.

- ✓ Sequence and Series.

- ✓ Statistics among others.

In Addition, each Chapter is accompanied with definitions and worked examples which create room for quick understanding. Timed revision exercises based directly on the work similar to those set in the said examinations are provided with the **Answers** at the end of the text to enable students test their ability.

Hence, provide opportunity for the additional practice that is needed for ensured success.

According to Albert Einstein ***"Make everything as simple as possible but not simpler".*** This Simplified Mathematics Series 1 for Secondary Schools with Answers has been devoted to developing deeper mastery of core topics as stated above through computational, problem solving and logical reasoning. It is meant to be "teach yourself" and "self explanatory"

We concentrated on how Mathematical instructions should start because having students' mathematical learning commence on the right paths is critical to all future mathematical learning in Schools and Colleges.

However most of the principles are relevant to Mathematics taught in all levels.

We had made all terms definable with languages that are mathematically accurate with key theorems, formulae and proofs wherever.

Finally, We recommend this book as a companion to all Mathematics preparatory students, aptitude testing candidates, Mathematics beginners and enthusiasts.

Chapter One

INDICES, EXPONENTS AND LOGARITHMS

INDICES: A number that indicates a characteristics or function in a mathematical expression is called an index. An index also is a number of or other symbol that indicate the location of a variable in a list or other mathematical objects.

In Algebra ,the word index sometimes denotes an exponent or the degree of an nth root. An exponent index is generally written as a superscript and an nth root index as a small numerical within the radical symbol. Example 2^5,the index is 5, and the index $^{1/3}$ in the expression $\sqrt[3]{8} = 8^{1/3}$ in a simplified manner.

For instance, in y^4 the exponent 4 is also known as the index. Similarly in $\sqrt[3]{27}$ and $\log_{10} x$ the number 3 and 10 respectively are called indices (indexes).

However, it is a number indicating a typical characteristic of an expression. For instance, in x^3, 3 is the exponent in the index, and in $\log_e{}^X$, e is the index.

EXPONENT: Power to which a number or expression is raised is called an exponent. This is written in the form of a superscript.

For instance, n and x are the exponents in a^n and $(ay + d)^x$

1.01 **PROPERTIES OF EXPONENTS**:

Multiplication : $X^a X^b = X^{a+b}$

Division : $X^a / X^b = X^{a-b}$

Power of Power : $(X^a)^b = X^{ab}$

Negative Power : $X^{-a} = 1/X^{\{a\}}$

Fractional Power : $X^{a/b} = X^{a-b}$

1.02 <u>FUNDAMENTAL LAWS OF INDICES AND EXPONENTS</u>

(i) $M^x \times M^k = M^{x+k}$

(ii) $(M^x)^k = M^{xk}$

(iii) $M^k \div M^x = M^{x-k}$ if k > x

   and $M^{-(x-k)}$ if k < x

LAWS OF INDICES AND EXPONENTS

(i) $M^x \times M^k = M^{x+k}$

(ii) $(M^x)^k = M^{xk}$

(iii) $M^k \div M^x = M^{x-k}$ if k > x

   and $M^{-(x-k)}$ if k < x

(iv) $M^0 = 1$

(v) $M^{-x} = \dfrac{1}{M^x}$

(vi) $M^{1/k} = \sqrt[k]{M}$

(vii) $M^{x/k} = \left(\sqrt[k]{M}^{\,x} \right)$

1.03 <u>APPLICATION OF INDICES AND EXPONENTS</u>

(i) $2^2 \times 2^3 = 2^{2+3}$ $= 2^5$

$$= 2 \times 2 \times 2 \times 2 \times 2$$

$$= \mathbf{32}$$

(ii) $(2^2)^3 \times 2^{(2+3)}$ $= 2^6$

$$= 2 \times 2 \times 2 \times 2 \times 2 \times 2$$

$$= \mathbf{64}$$

(iii) $2^3 \div 2^2 = 2^{3-2}$ $= 2^1$

$$= 2 \ (\text{ If } 3 > 2 \ )$$

and $2^2 \div 2^3$ $= 2^{2-3} = 2^{-1}$

$$= \mathbf{\frac{1}{2}} \ (\text{ If } 3 < 2 \ )$$

(iv) $2^0 = 1$

(v) $2^{-2} = \frac{1}{2}^2 = \frac{1}{4}$

(vi} $2^{1/3} = \sqrt[3]{2}$

(vii) $2^2{}_{/3} = \left(\sqrt[3]{2}\right)^2$

EXAMPLE 1.00:

1. Given $8^{-2/3}$

$$= \frac{1}{8}^{2/3}$$

But $8^{2/3} = \left(\sqrt[3]{8}\right)^2$

$\qquad = (2)^2 = 2 \times 2 = 4$

$\qquad 1/8^{2/3} = \frac{1}{4}$

2. $(0.3)^{-2}$

$\qquad = 1/(0.3)^2$

But $0.3 \qquad = 3/10$

$\qquad 1/(3/10)^2 = 1 \div (3/10)^2$

$\qquad = 1/1 \times (3/10)^2$

$\qquad = 1/1 \times 100/9$

$\qquad = 100/9$

REVISION EXERCISE 1.00:

1. $(0.4)^2$

2. $(64/27)^{-2/3}$

1.04 LOGARITHM THEORY AND APPLICATIONS

This can be referred as a number which may be expressed as the exponent { *power of another number called its base*}

It is an extension of indices. They are closely related, for instance

$1 = 10^0$

$10 = 10^1$

$100 = 10^2$ etc

And,

$3 = 3^1$

$9 = 3^2$

$27 = 3^3$

$49 = 7^2$ etc.

However, $\log_{10} 1 = 0$

$\log_{10} 10 = 1$

$\log_{10} 100 = 2$

$\log_{10} 10000 = 3$ etc

And,

$\text{Log}_3 1 = 0$

$\text{Log}_3 3 = 1$

$\text{Log}_3 9 = 2$

$\text{Log}_3 27 = 3$

$\text{Log}_3 81 = 4$ etc

So the logarithm of n number to m base is the power or index to which the base, m must be raised to give that number, n.

EXAMPLE 1.02:

$8 = 2^3$ (*Index form where 2 equals base and 3 equals index*) so, $Log_2 8$ is read as log of 8 to base 2

Meaning, $Log_2 8$ $= Log_2 2^3$

$$= 3Log_2 2$$

$$= 3 \times 1 = \mathbf{3} \quad (since\ Log_2 2 = 1\ or\ Log_n n = 1)$$

Where n = non negative number

EXAMPLE 1.03: Solve $Log_{10} 100 = x$

$= Log_{10} 100 = Log_{10} 10^2$

$= 2Log_{10} 10 \qquad$ but $Log_{10} 10 = 1$

So $\quad 2 Log_{10} 10 = 2 \times 1 = \mathbf{2}$

EXAMPLE 1.04:

Solve for x, if $Log_{10} 0.01 = x$

SOLUTION

$$10^x = \frac{1}{100}$$

$$= \frac{1}{10^2}$$

$$= 10^{-2}$$

$$10^x = 10^{-2} \qquad \textit{(Same base)}$$

$$x = \textbf{-2}$$

REVISION EXERCISE 1.01:

1. $\text{Log}_{7-1} 49 = x$

2. $\text{Log}_5 25 = x$

LAWS OF LOGARITHMS

Similar to the three (3) fundamental laws of indices are three (3) fundamental laws of logarithms thus:

$$\text{Log MN} = \text{Log M} + \text{Log N} \qquad\qquad \text{equ (i)}$$

$$\text{Log M} = \text{Log M} - \text{Log N} \qquad\qquad \text{equ (ii)}$$

$$\text{Log (M)}^N = N \log M \qquad\qquad \text{equ (iii)}$$

Let $\quad \text{Log}_a M = q$ and $\text{Log}_a N = r$

where "a" is the base

$$a^q = M \text{ and } a^r = N$$

So $\quad \dfrac{M}{N} \quad = \dfrac{a^q}{a^r}$

$\qquad = a^q - a^r$

$Log_a \left[\dfrac{M}{N}\right] \quad = q - r$

$\qquad\qquad = Log_a\ M - Log_a\ N$

Having "a" as common base, can take any value

So, $Log\left[\dfrac{M}{N}\right] = Log\ M - Log\ N$ true for all given base

1.05 LOG OF NUMBERS {GREATER THAN ONE AND LESS THAN ONE}

LOGARITHM OF NUMBERS GREATER THAN ONE

Applying the laws of indices, we simplify 3.483 x 54.27 using the logarithm table as follows;

EXAMPLE 1.05:

$$3.483 \text{ x } 54.27 \quad = 10^{0.5420} \text{ x } 10^{1.7346}$$

$$= 10^{0.5420\,+\,1.7346}$$

$$= 10^{2.2766}$$

From the antilogarithm table, we look for .27 under 6 add difference 6, thus;

```
    1888
+      3
   1891
```

Having the characteristics as 2, we move $(2 + 1) = (3)$ three places of decimals to the right, thus; 1891 becomes 189.1

$$3.483 \times 54.27 = \textbf{189.1}$$

EXAMPLE 1.06:

$$67.52 \div 35.81 \quad = 10^{1.8294} \div 10^{1.5541}$$

$$= 10^{(1.8294 - 1.5544)} \quad (\textit{Law of Indices \& Log applied})$$

$$= 10^{0.2753}$$

Finding the antilogarithm of 27 under 5, add difference 3 thus

```
    1888
+      3
   1885
```

Having the characteristics as zero, (0), we move $0 + 1 = (1)$ one place of decimal to the right, Thus 0.1885 becomes 1.885

$$67.57 \div 35.81 = \textbf{1.885}$$

REVISION EXERCISE 1.02:

1. 40.9 x 69.32

LOGARITHM OF NUMBERS LESS THAN ONE

As earlier noted in finding logarithm of numbers greater than (1) one that the characteristics are the power to which (10) ten is exponentials to the non-negative (integers) numbers in standard forms.

However, the status quo characteristics are found by considering the negative powers to which (10) ten is exponentials in standard forms.

EXAMPLE 1.07:

$$\frac{0.987 \times 0.864}{0.753}$$

Having 0.987 x 0.864 as

No	Log
0.987	.9943
0.864	.9365 (+)
	.9308
	.9308
0.753	.8768 (-)
	0.0540
	1164

Finding the antilog of 0.0540

➔ Antilogarithm of .05 under 4, add difference of zero (0)

i.e.   1132
    +   0
    ‾‾‾‾‾‾‾‾‾
       1132

Having the characteristics as zero we move 0+1 =1 place to the right, thus becomes **1.132**

REVISION EXERCISE 1.03:

1.

$$\sqrt{\frac{\{\ 3.68\ \}^2 \times 6.705}{0.3581}}$$

2.

$$\sqrt{\frac{0.897 \times 3.536}{0.00249}}$$

APPLICATION OF THE THEORY

1. $Log_4 X = 64$ find X

   $Log_4 X = 64$

   $4^{64} = X$

2. $Log_x 8 = 3$

   $x^3 = 8 = 2^3$

   $x^3 = 2^3$ (*Having the same indices*)

   $x = 2$

3. $Log_x 81 = 2$

   $x^2 = 81 = 9^2$

   $x^2 = 9^2$ (*Having the same indices*)

   $x = 9$

4. $8x^{-2} = {}^{2}/_{25}$

   Dividing through by 2

   $$\frac{8x^{-2}}{2} = \frac{2}{25} \div \frac{2}{1}$$

   $$\frac{8x^{-2}}{2} = \frac{2}{25} \times \frac{1}{2}$$

   $4x^{-2} = {}^{1}/_{25}$

   $4x^{-2} = (25)^{-1}$

   $4x^{-2} = 5^{2(-1)}$

   $4x^{-2} = 5^{-1}$

   $$\frac{4 \times 1}{x^2} = \frac{1}{5^2}$$

Cross multiplying

$$4 \times 1 \times 5^2 = x^2 \times 1$$

$$x^2 = 100$$

$$x = \sqrt{100} = 10$$

Chapter Two

SET THEORY

This is a well defined collection of objects or things.

EXAMPLE 2.00: Set of prime numbers less than 10 are

(1, 3, 5, 7). This is usually represented by the symbol { } or by the use of capital letter of the alphabets.

2.01 SET AND ITS ALGEBRAIC NOTATIONS

ELEMENTS OF A SET (ε): This implies the set of items belonging to the set. It is also known as members of a set.

Moreover {ε̸} implies "not an element"

EXAMPLE 2.01: If Q is { 1, 3, 4, 9 }

It implies that 3 ε Q and 5 ε̸ Q meaning that 3 is an element of set Q and 5 is not an element of the set Q respectively.

NUMBER OF ELEMENTS IN A SET: This denotes the number of the elements that are found in that Set.

EXAMPLE 2.03: If M = { 2, 3, 4, 6, 5, 8, d }

i.e n{ M} = 7

SUBSET: This is a set within another Set, also is a set contained in another set.

EXAMPLE 2.04: Let x = {3, 5, 4, 9, 12} and y = {5, 9, 12}

∴ $y \subseteq x$

y = proper subset of x, because it contain not all the element of x.

UNIVERSAL SET (μ): This is set contains all the items of all its subsets.

POWER SET: This is the number of subject which a particular set has.

EXAMPLE 2.05: If P = {4, 3}

Subset = { } {4} {3} {4, 3}

Also, if Z = { 3 }

Subset = (3) and { }

∴ Power set = 2^n

where n= number of elements

EXAMPLE 2.06:

Find the power set of the following

(a) T = {prime factors of 210 }

SOLUTION:

Since it implies finding the Lowest Common Multiple of 210

thus,

1	210
2	105
5	7
7	7
	1

$$\therefore 2 \times 3 \times 5 \times 7 = 210$$

The prime factors of 210 are:

(2, 3, 5, 7)

But the power set is 2^n by formula,

where n = 4

$$\therefore 2^n = 2^4$$

$$= 2 \times 2 \times 2 \times 2$$

$$= \mathbf{16}$$

ALGEBRAIC NOTATIONS

EXAMPLE 2.07:

Q = { x : x is a prime number less than 30 }

Interpreting the above,

Q = { 2, 3, 5, 7, 11, 13, 17, 19, 23, 27, 29 }

EXAMPLE 2.08:

M = { x : x ε Z, 3 ≤ x ≤ 10 },

Interpreting,

M= { 3, 4, 5, 6, 7, 8, 9, 10 }

So   **x ≥ 3 ; x ≤ 10**

2.02 SETS AND ITS RELATIONSHIP

1. **EMPTY SET:** This is a set without any element or member. It is also called null set with a notation as " ø " or { }

   NOTE: Zero (0) is not an empty set rather it is an element of a set.

2) **FINITE SET:** This is countable element of a set. It shows element having definite end.

3) INFINITE SET: This is uncountable element of a set

4) DISJOINT SET: This involves two sets having no element in common.

5) EQUAL SET: These are sets of the same element irrespective of the arrangement.

EXAMPLE 2.09:

If   Q = (3, 4, 1, 8) and P = (4, 8, 1, 3)

Hence,   Q = P

SET RELATIONSHIP

UNION OF SET: This is the set of all elements in two or more sets together as one. Denoted as "U"

EXAMPLE 2.10: If M = { a, b, c, 2, d} and Q = { 3, 4, a, c, b}

Then MUQ = { a, b, c, d, 2, 3, 4 }

DISJOINT SET: Two sets having no element in common is called a disjoint set

EXAMPLE 2.11:

If M={ 4, 3, a, j,f } and K={ 1, 5, z , q } then M and K are disjoint.

EXAMPLE 2.12

Given the set of numbers such that Q = { 5, 6, 7, 8 } and H = { a, b, c, d } then Q and H are disjoints.

INTERSECTION OF SETS:

This is a set of all the elements that are common to both sets.

Denoted as "∩"

EXAMPLE 2.13: If M = { a, b, c, 2, d}

and   K = { 4, 5, c, d}

Then M∩K = { b, c, d}

DIFFERENCE OF SET: This implies the element of a set that is not found in another set. Denoted as A-B

EXAMPLE 2.14: Given A = { 3, 6, 8, 10} and

$\qquad$ B = { 6, 8, 10}

∴   A-B = **{ 3 }**

It is read as "A" difference "B" or A/B

So, symmetric difference denoted as $\triangle$ is the union of the difference in the two sets, when separately determined, thus,

$$A \triangle B = A/B \; \mathbf{U} \; B/A$$

COMPLEMENT OF A SET:

This is the set of all elements which are found in the Universal set $\{\mu\}$ but are not found in the set itself. Denoted as $\{\,'\,\}$ or $\{\,^c\,\}$

EXAMPLE 2.15: If $\mu = \{\,a, b, c, d, e\,\}$

L = $\{\,a, c, d\,\}$ then

$L^c = \{\,b, e\,\}$ and

B= $\{\,b, e\,\}$ then

$B^c = \{\,\mathbf{a, c, d}\,\}$

SUPERSET: Is a set that contain its subject property. Denoted as ">"

EXAMPLE 2.16: If A = $\{\,1, 2, 3\,\}$

B = $\{\,1, 2, 3, 4, 5\,\}$

Then B > A

i.e. B is a superset (or supersedes) of A because B contains A property.

2.03 <u>VENN DIAGRAMS AND PROOFS</u>

This is the diagrammatical representation of sets relationships enclosed in a rectangle with circular representation of elements of each subset in a given set. It is best used in solving complex problems.

(1) UNION SET: MUQ ={ a, b, c, d, 2, 3}

If M = { **a, b**, c, 2, d}

    Q = { 3, 4, **a**, c, **b**}

Diagrammatically

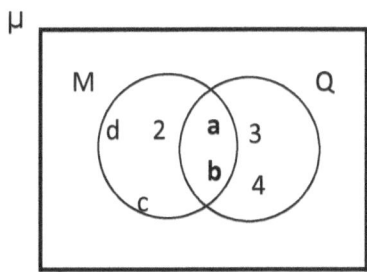

(2) DISJOINT SET

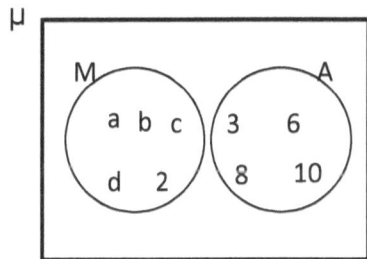

ie No intersection

Implying that   M∩A = Ø

Where   Ø = Null

(3) INTERSECTION

If   M = { a, **b**, **c**, 2,**d**}

       K = {**b**, **d**, **c**, 4}

       M∩K = (b, c, d}

Diagrammatically

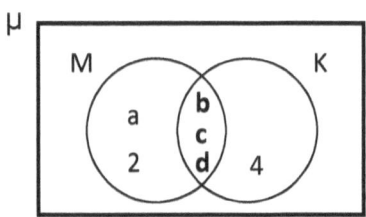

4) DIFFERENCE OF SETS

If   A = { 3, 6, 8, 10 }

       B = { 6, 8,10 }

       A -B = { 3 }

         = A ∩ Bᶜ   where Bᶜ = { 3 }

However, μ = { 3, 6, 8, 10 }

5) COMPLEMENT OF A SET

Diagrammatically

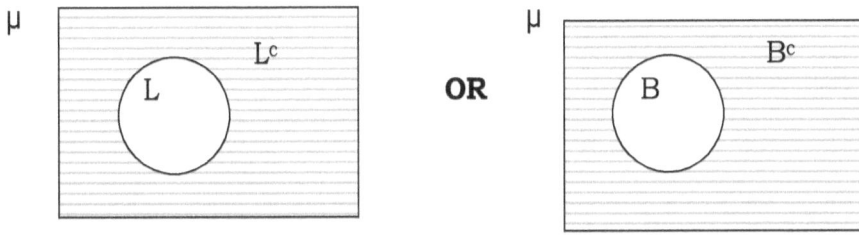

OR

{6} SUPER SET

Diagrammatically

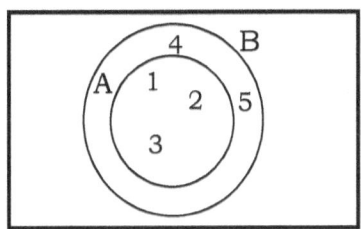

HARDER RELATIONSHIP (THREE SETS: P,Q,R)

EXAMPLE 2.13:

1) UNION SET:

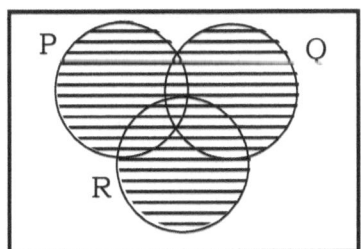

2) DISJOINT SET:

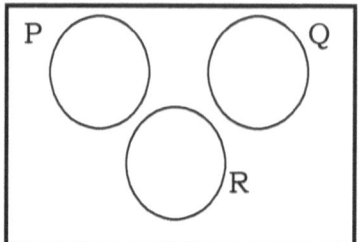

3) INTERSECTION (THE SHADED PORTION)

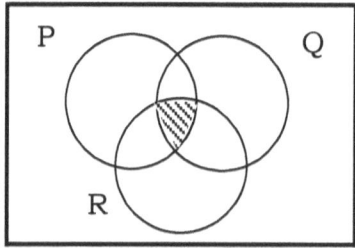

P∩Q∩R

Also,

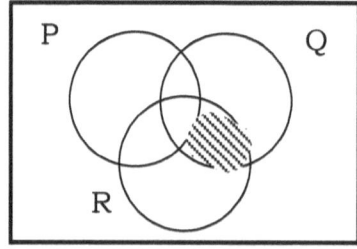

Q∩R∩P^c i.e the shaded portion

However,

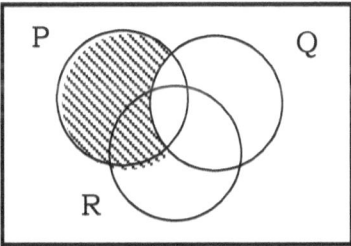

$P \cap Q^c \cap R^c$ i.e the shaded portion meaning element in P not in Q and not in R.

SET PROBLEMS

1) In a class of 50 students, 4 students do not study Mathematics or Physics; if 40 students study Mathematics and 32 students study Physics, how many students study Mathematics alone?

SOLUTION

Diagrammatically with a Venn diagram

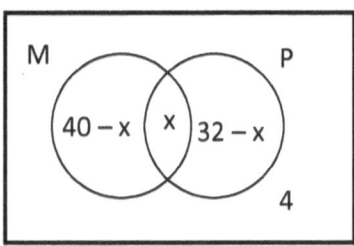

$40 - x + x + 32 - x + 4 = 50$

$76 - x = 50$

Bringing like terms together

76-50 = x

∴ x = 26

40 - x + x + 32 - x = 50

76 - x = 50

: . 26 = x

x = 26

However, the number of students studying Mathematics is

40 - x = 40-26 = 14

REVISION EXERCISE 2.00:

1. In a science class of 42 students, each student offers at least one of Mathematics and Physics. If 28 offers Mathematics, 22 offer Physics, find how many students that offers Physics only.

Chapter Three

COMMERCIAL / FINANCIAL ARITHMETICS

3.01 SIMPLE INTEREST (S.I)

This is expected charge from the borrower as compensation to the lender when money is borrowed.

This has its formula as S.I = $\frac{P \times T \times R}{100}$ equation.

where   P = Principal (the original amount lent out)

               T = (the period in years, month, weeks etc for which the money is borrowed)

               R = (the amount of compensation i.e. interest as percentage of the amount lent out; principal: original amount)

Nevertheless, each of the above variables can be made the subject of the formula to find the others, thus

$$I = \frac{PTR}{100} \qquad So, P = \frac{100I}{TR}, \qquad T = \frac{100I}{PR} \text{ and } R = \frac{100I}{PT}$$

EXAMPLE 3.00: Find the simple interest, (S.I) on ₦2400 in 8 years at 5½% per annum.

SOLUTION

S.I = $\dfrac{P \times T \times R}{100}$, But P = ₦2400, R = 5½% = 5.5% and T = 8years

$\quad$ S.I = $\dfrac{2400 \times 8 \times 5.5}{100}$ = **₦1,056**

3.02 COMPOUND INTEREST:

This is the interest earned on capital where the interest is periodically added to the principal. The formula for calculating the final Amount, A when the interest has been added to the principal n times is

A = $P \left(1 + \dfrac{r}{100}\right)^n$ $\qquad\qquad$ equation.

Where A $\quad$ = A_f

$\qquad\qquad$ = final Amount

$\quad$ P $\quad$ = Principal/original amount

$\qquad\qquad$ = $A_{original}$

$\quad$ n $\quad$ = number of years

$\quad$ r $\quad$ = rate/ Return rate.

By formula, A = P + I

$\Rightarrow \quad A_{final} = A_{original} + I_{compd}$

However instead of using the long process of substituting for n = 1, 2, 3 . . . etc, the above method will ease work and save time.

EXAMPLE 3.01: Find the Compound Interest, (I_{compd}) on ₦9000 in 3 years at 4% per annum.

SOLUTION (Long process)

If $A_{final} = P(1 + {}^r/_{100})^n$, $I_{compd} = A_{final} - A_{original}\{P\}$

For the n = 1, $I_1 = \dfrac{P \times T_1 \times R}{100} = \dfrac{9000 \times 1 \times 4}{100} = ₦360 \div 1 = ₦360$

For n = 2, Principal, P = 9000 + 360 = 9360

$\therefore I_2 = \dfrac{P \times T_2 \times R}{100} = \dfrac{9360 \times 2 \times 4}{100} = 748.2 \div 2 = 374.4$

$I_3 = \dfrac{P \times I_3 \times R}{100} = \dfrac{9734.4 \times 3 \times 4}{100} = 1168.128 \div 3 = 389.376$

$= 374.4 = 9734.4$

N.B: Final A_f = ₦9734.4 + 389.376 = 10123.776

Original Amt = Principal = ₦900

$\therefore I_{compd} = A_f - A_{orig} = 10123.776 - 9000 = ₦1123.776$

FORMULA METHOD OF SOLVING COMPOUND INTEREST

EXAMPLE 3.02: Find the Compound Interest of N9000 in 3 years at 4% per annum.

SOLUTION: P = N9000 = $A_{original}$ Original Amt. n = 3yrs r = 4%

By formula, $A_{final} = P \left(1 + {}^{r}/_{100}\right)^n = 9000 \left(1 + {}^{4}/_{100}\right)^3$

$= 9000 \times (104 \div 100)^3 = 10123.7760$

$\therefore A_{final} = ₦10123.7760$

But Compound Interest $= A_{final} - A_{original} = 10123.7760 - 9000$
$= ₦1123.776$

EXAMPLE 3.03: A customer deposited N3570.00 in a fixed deposit account for 4yrs at 12% per annum. Determine the Interest paid at the end of the 4th year.

SOLUTION

$P = A_{final} = ₦3570$, r = 12% p.a, n= 4yrs

$\therefore A_{final} = P \left\{ 1 + {}^{r}/_{100} \right\}^n$

$I_{compd} = A_{final} - A_{original}$

$I_{compd} = A_{final} - P = 3570\left\{ 1 + {}^{12}/_{100} \right\}^4$

$= 3570 \times \{ 112 \div 100 \}^4 = 3570 \times \{ 1.12 \}^4 = 5617.45 \Rightarrow A_{final}$

$\therefore I_{compd} = A_f - A_{original} = 5617.46 - 3570 = ₦2047.49$

3.03 <u>DISCOUNT</u>

This is the difference between the issue price of a stock of share and its normal value when the issue price is less than the normal value. This is also allowance in percentage allowing a buyer by the seller or the purchase price an article.

TYPES OF DISCOUNT

1. **CASH DISCOUNT:** is the allowance given to the buyer by the seller for payment in cash for articles bought either immediately or within a stipulated time frame/ limit.

2. **QUANTITY DISCOUNT:** is the allowance that allowed the buyer for buying articles at stipulated quantity.

EXAMPLE 3.04:

A N200 Note drawn up on 9th January 2009 for 100 days is deposited at a bank on February 15th 2009. The bank charges a 6.5% discount on Notes. How much will the depositor receive?

SOLUTION

From 9thJanuary -15th February ⇒ 36days. i.e. the Note has 64days to run

{ *100 days minus 36days* }

$\Rightarrow$ 64days = 64day/360days $\{$ *of the year* $\}$ $\Rightarrow$ Time

T = 0.1777, R = 6.5, Principal P = N2000

Interest = $\dfrac{P \times R \times T}{100}$ = $\dfrac{2000 \times 0.1777 \times 6.5}{100}$ = 23.101

:. The depositor will receive, P - Interest

=N2000 - N23.101= **₦1976.89**

EXAMPLE 3.05:

After allowing a discount of 9% on an article, a seller collected the sum of ₦1865.50 find (a) the total sales/ Marked price (b) the Amount of Commission paid.

SOLUTION

(a) Discount = 9%, Percentage paid = $\{$ 100 - 9 $\}$ % = 91%, Marked price = $\{$1865.5 /91%$\}$ = 1865.5 $\div$ $^{91}/_{100}$ = 1865.5 x $\{$ $^{100}/_{91}$$\}$ = **2050** $\Rightarrow$ Total sales.

(b) Percentage commission = 9%, amount involved = ₦2050

:. Amount of commission = 9% of the amount involved.

So,$\{$ percentage commission$\}$ X$\{$ the amount involved $\}$ =

$\{$ 9 $\div$100 $\}$ X $\{$ 2050 $\div$ 1 $\}$ = **₦184.50**

Discount = 6%

Percent paid = 94% = { 100 - 6 }% = ₦846

: . 94% = N846, So MP = { 846/94 } x { $^{100}/_1$ } = ₦900

(b) Cost price { C.P } to the seller is ₦750

(c) To find discount allowed, knowing that 6% of 900 = ₦54 or discount is MP - Amount sold ⇒ ₦{ 900 - 846 } = **₦54**

EXAMPLE 3.06:

Find the price of an Article if the ₦60.50 Commission allured is 5%

Discount 5%

=6050

Marked Price, MP = $\dfrac{60.50}{5} \times \dfrac{100}{1}$

= ₦1200

EXAMPLE 3.07:

An agent collected an article ₦40 Commission on sales made on article marked ₦1600. What is the rate of the Commission in

(i) Percent ?

(ii) In kobo per ₦ ?

SOLUTION

Percentage Commission = x %

Amount involved = ₦1600

∴ Amount of Commission , ₦40 =

$\quad$ x% of 1600

$\quad$ $40 = {}^x/_{100}$ of 1600

∴ $\quad$ $40 = \dfrac{1600}{100}$

$\quad$ $40 = 16x$; $x = {}^{40}/_{16} = 2.5\%$

$\quad$ **= 2.5k**

EXAMPLE 3.08: A debtor is allowed a discount of 10% if he pays before the end of the month and he paid the money. The amount owned is ₦150000.How much is the he going to pay?

SOLUTION:

If 10% of ₦150000 = ₦15000,

i.e $^{10}/_{100}$ x ₦150000 = ₦15000

Then the amount to be paid = ₦{ 150000 - 15000 }

$\quad$ **= ₦135000.**

NOTE: If a discount is mmade ,then the remaining payment is called the percentage paid.

EXAMPLE 3.09: If I discount 6% from a particular goods, then the percentage paid is { 100-6} = **91%**.

SELLING PRICE (S.P) AND COST PRICE (C.P)

EXAMPLE 3.10: if the Cost Price of an article is N100 and the profit is N10. Find the Selling Price. { S.P }

S.P　= C.P + Profit

= 100 + 10` = **110**

From the above equation either C.P or S.P can be determined by the method of subject of the formula.

We also know that S.P = C.P -loss

:. Loss = C.P -S.P

FINDING S.P IF PROFIT IS EXPRESSED BASED ON C.P

EXAMPLE 3.11: Find the S.P of an article at the profit of 20% of N500 Cost Price.

SOLUTION

First: % of profit x C.P

= 20% x 500

= $^{20}/_{100}$ x 500

= ₦100 (in terms of money)

Second: ₦500 + ₦100 = **₦600**

FINDING S.P IF LOSS IS EXPRESSED BASED ON C.P

EXAMPLE 3.12: Find the S.P of an article at the loss of 20% of the Cost Price ₦500 or article cost ₦500 and sold at the cost of 20%, find the S.P

SOLUTION

First: % of loss x C.P

$= 20\% \times 500$

$= \frac{20}{100} \times 500$

₦100 { in terms of money }

Second: ₦500 - ₦100 = **₦400**

PERCENTAGE PROFIT BASED ON C.P

EXAMPLE 3.13: Find the percentage of profit based on C.P of an article costing ₦3.50k and selling for ₦4.00

SOLUTION

C.P = 3.5k

S.P = 400

Profit = C.P - S.P

= 0.50

∴. $\dfrac{0.50}{3.50}$ = **0.1428**

PERCENTAGE LOSS BASED ON C.P

EXAMPLE 3.14: Find the percentage of loss on C.P of an article costing ₦4.00 and selling for ₦3.5

SOLUTION

C.P

S.P N3.50

Loss = C.P -S.P = 0.5

∴. $\dfrac{0.5}{4.00}$ = 0.125 = **12.5%**

FINDING C.P WHEN GIVEN THE S.P AND THE PERCENTAGE PROFIT BASED ON THE S.P

EXAMPLE 3.15: If an article sells for ₦25 and there exist a profit of 10% of the S.P, Calculate the C.P

SOLUTION

S.P =₦25

Profit = ₦25 x 10%

= $\frac{25 \times 10}{100}$ = 2.5

Then S.P - profit

= ₦25 - ₦2.5

= **22.5**

FINDING C.P WHEN GIVEN THE S.P AND THE PERCENTAGE LOSS BASED ON THE S.P

EXAMPLE 3.16: Assuming an article sells at ₦25 and there exist a loss of 10% of the S.P Calculate the C.P

SOLUTION: S.P = ₦25

Loss { in terms of money } = ₦25 x 10%

= 25 x $\frac{10}{100}$

= 2.5

= ₦25 + ₦2.5 = **₦27.5**

But C.P = S.P + loss

TAX AND EXCHANGE RATE

TAX: Is a compulsory payment made by citizens and corporate bodies to the government to enable it carry out its responsibilities. Everybody who earns an income above a certain minimum amount is expected to pay part of the income as income tax.

PAY-AS-YOU-EARN (P.A.Y.E)

This applies to income tax deducted and paid according to the payee's income and responsibilities. Here, the amount of tax deductible depends on the employee's total pay and the employee's income tax allowances. Where as a progressive tax entails tax on higher income at higher rate and lower income at lower rate.

Allowances are set against income tax such as personal allowances, child allowances, dependant relatives, premiums paid on social scheme-insurance, charity organizations etc.

However, all deductions are made from the Gross Income {G.I} while others {Taxable *Income*} are taxed with respect to different rate payable on large amount of income.

3.04 <u>INVESTMENTS</u>

(i) INVESTMENT

This is a way of saving money. This is done by either depositing a certain amount of money in a bank or by buying shares of a company.

In commercial Banks, we have:

(ii) NORMAL SAVINGS ACCOUNT: Here the customer is allowed to withdraw or add the principal amount deposited.

(iii) FIXED DEPOSIT ACCOUNT: Here the customer enters into an agreement with the bank for a fixed period of time during which he/she cannot withdraw any part of the savings and interest as fixed deposits is paid at the end of the agreed period only after which he/she can withdraw the whole amount or re-invest part or whole.

However, all savings in commercial banks attracts compound interest thus:

P = Amount Invested

R = at the rate of

n = Per annum (years)

Then total Amount (A) = $P (1 + R/100)^n$

3.05 <u>SHARES AND STOCKS</u>

Private individuals find it not easy to provide sufficient funds or capital for his business for it to be a public company.

In any public company, shares are issued and public are invested to subscribe {*buy*} for these shares-normally issued in units at 10k, 50k, ₦1 etc.

The unit implies to Nominal value of the shares of which at end of every financial year, the company declares a dividend (benefit) in the form of percentage (%) of the Nominal value.

And each share-holder paid the dividend due on his holding, however, shares could also be sold at a discount meaning selling it at amount lower than Nominal value. While stock can be subdivided, shares cannot be subdivided.

EXAMPLE 3.17: A dividend of 15% means 15k for each ₦100 share or 7.5k for each 50k share

CONTROL OF SHARES

The buying and selling of shares is controlled by Stock Exchange. Good dividend means good deeds by a company showing increase in the price & shares which goes up, but if the company does not do well, then the quoted price of the shares goes down.

However, fluctuation in the price of the shares (*does not implies*) nor change the Nominal value (*constant*). The Stock Exchange price for shares is quoted in the daily News papers and Televisions.

TYPES OF SHARES

{1} **PREFERENCES SHARES**-Is issued at a fixed percentage. The payment of its dividend is the first charge on the profit of the company. This is safe but not likely to appreciate or depreciate capital value as in ordinary share.

{2} **ORDINARY SHARES**-The dividend is not fixed and declares by the Directors of the company in line with the profit made, thus, it preference dividend takes all or almost all the profit of the company, then the ordinary shareholders may get nothing.

However, when there is bumper year, the ordinary shareholder may get a much large dividend.

EXAMPLE 3.18: A shareholder purchased 4000 shares of Nominal value of 25k each at 32kbo and a dividend of 14% is dividend. Find the income derived that is value of the dividend.

SOLUTION

{Value of dividend in % of declared} x{ shares bought} x (Amt of Nominal value)

$$= \frac{14}{100} \times \frac{4000}{1} \times \frac{0.25k}{1}$$

$$= 0.14 \times 4000 \times 0.25k$$

$$= ₦140$$

EXAMPLE 3.19: From 9.5% preference shares of Nominal value ₦1.00 quoted at ₦0.92, Calculate the dividend expected from an investment of ₦3680.00

Amount of Nominal (dividend declared) value = 0.92

Amount invested = ₦3680.00

$$: . \text{ Number of shares} = \frac{3680}{0.92}$$

$$= 4000 \text{ shares}$$

: . Dividend of 9.5% preferences share of Nominal value

$$= \frac{9.5}{100} \times \frac{4000}{1} = ₦380$$

$$= ₦380 \times ₦0.92k \quad (Amt \text{ of Nominal Value})$$

= ₦349.60k

EXAMPLE 3.20: Find the yield on a 25k share quoted at 32k when the dividend is 18%

SOLUTION

A dividend of 18% on 25k $= \dfrac{18}{100} \times \dfrac{25}{1}$

$$= 4.5k$$

$: . \% \text{ yield} = \dfrac{4.50}{32} =$ **14.06%**

EXAMPLE 3.21:

Calculate the yield on a 50k share quoted at ₦0.45 when the dividend is 4½%

SOLUTION

    4.5% of 50k

    $\frac{4.5}{100} \times \frac{50}{1} =$ **2.25k**

EXAMPLE 3.22: A share of Nominal value 10 is quoted at ₦14k and pays income on investment of ₦2800.00 cash. 10% dividend will be calculated as ;

SOLUTION

Amount of Nominal value = ₦0.10k

    ∴ Amt invested = ₦2800

Number of shares = $\frac{2,800}{0.10k}$ = 2800

∴ Dividend of 10% = $\frac{10}{100} \times \frac{2800}{1} \times \frac{0.10}{1}$ (Amt of Nominal value)

        = **₦280.00**

NOTE: Nominal value = dividend declared

Meaning that Nominal value = Dividend

: . Nominal value = Selling Price (*of the shares*) - share (*value*)

EXAMPLE 3.23: If ₦5000 valued at ₦1.00 share is sold for ₦1.50k to an investor. What is the dividend declared or Nominal value amount?

SOLUTION

Dividend declared = Selling Price (*of the share*) - share value

    = ₦(1.50-1.00)

    = **₦0.50k**

3.06 ANNUITY

This is a fixed amount of money paid annually for the rest of someone's life. It is a form of insurance income paid at regular interests either directly to the beneficiary or through a savings account. It is yearly allowance provided by an investment.

EXAMPLE 3.24: If a man has an annuity of ₦1000, then he will be receiving ₦1000 every year, either directly or through this savings account. At the end he/she will collect Annuity plus Compound interest.

3.07 <u>SOME KEY WORDS IN FINANCIAL ARITHMETICS</u>

DIVIDEND: Is share of profits payable. It is also benefit from an action.

GRATUITY: Is money given as a present for service rendered.

GRANTS: Given or allowed as a privilege. Admit to be true students allowance from public funds.

ALLOW: Permit or giving a limited quantity or sum.

ALLOWANCE: Amount or sum allowed

GROSS INCOME: Total income before tax

TAXABLE INCOME: Gross income less Allowance

TAX PAYABLE: Total amount deducted from taxable income (mainly calculated from the information given by the examiner)

3.08 <u>CALCULATIONS ON TAXATION AND COMMISION</u>

EXAMPLE 3.25: How much will be realized on ₦4000 if tax rate is at 15%.

SOLUTION

Tax rate = 15% = 0.15%

Base = ₦4000

: . Tax = Tax rate x Base

   = 0.15 x 4000

   = **₦600**

PROBLEMS WHEN GIVEN THE BASE, TAX RATE IN TERMS OF MONEY

Given tax rate = ₦360 per ₦1000

Find the tax on ₦470, 500

SOLUTION

$$\frac{470,500}{1000} = 470.5$$

: . 470.5 x ₦360 = **₦1,693.80**

EXAMPLE 3.26: Solve a problem here please?

COMMISSION

This is a fee charged for transacting business for another person.

EXAMPLE 3.27: A sales man earns a salary of ₦200 weekly, plus a commission based on sales volume for the week.

The commission is 17% for the first ₦1500 and 10% for all sales in excess of ₦1500.

How much did he earn in a week in which his sales totaled ₦3200?.

SOLUTION

$$\begin{array}{r} ₦3,200 \\ - \underline{₦1,500} \\ \mathbf{₦1,700} \end{array}$$ meaning excess sales

7% = 0.07 x ₦1500 =	₦105	commission on 1st ₦1500
10% = 0.10 x ₦1700 =	₦170	commission on excess sales
	+	
	₦200	weekly sales
	₦475	**Total earnings**

EXAMPLE 3.28:

Mr. Okoko earns ₦550 salary per week with deduction as:

PHCN	= ₦106.70
FMB	= ₦41.31
State tax	= ₦22.83
Pension	= ₦6.41
Union Dues	= ₦5.84
	= **₦183.10**

Gross pay, G.P (Amount of money earned other from salary commission or both before any deduction are made)

= ₦550

Deductions = ₦183.10

: . Net pay (Take home pay) = G.P - deduction

= ₦(550-183.10)

= ₦366.90

INFLATION AND DEPRECIATION

INFLATION

This is the opposite of deflation. Inflation means loss in value of money.

It is the percentage increase in the cost of buying things from a particular year to another.

EXAMPLE 3.29: Given the rate of inflation as 20%, then a chair which cost ₦4000 a year ago will now cost;

SOLUTION

₦4000 x $\dfrac{120}{100}$ (*meaning 100% + 20%*)

= ₦4800 meaning that money has lost some of its value since it now cost more to buy the same thing.

EXAMPLE 3.30: In a certain year, a new cooking stove cost ₦16000. If the rate of inflation was 50% for the next two years, what would the same cooking stove cost at the end of the period?

SOLUTION

Initial cost $\qquad$ = 16000

Rise (1ST year costing) + 8,000   *(meaning ½ of 16000)*
$\qquad\qquad\qquad$ ₦24,000

After 1st year, cost $\qquad$ 24,000

Rise (2nd year cost) =+ 12,000   *(meaning ½ of 24,000)*
$\qquad\qquad\qquad$ ₦36,000

∴. After 2nd year cost, it will cost **₦36,000**

DEPRECIATION

This means loss in value of goods. It is the condition when item loss its value after being for a certain period of time. It is usually given in terms of percentage (%) of the value of the item every year.

EXAMPLE 3.31: A radio costing ₦16,000 depreciates by 25% in the first year and by 20% in the second year. Find its depreciative value after 2 years.

SOLUTION

25% depreciation of ₦16,000 = $\frac{25}{100} \times \frac{16,000}{1}$

$$= ₦4000$$

So, ₦(16,000 - 4,000) = ₦12,000 which in the depreciative value in the fist year.

Also 20% depreciation of ₦12,000 = $\frac{20}{100} \times \frac{12,000}{1}$

$$= ₦1,200$$

So, ₦(12,000 - 1,200) = **₦9,600** which is the depreciative value in the second year (also the answer).

However, in business education, depreciation can be defined as the reduction in the economic service potentials of an assets as a result of wears, tear usage and passage of time, depletion, inadequately and obsolescence.

ELEMENTS OF DEPRECIATION

ORIGINAL COSTS: This is cost incurred in purchasing, installation and cost carriage.

ESTIMATED/RESIDUAL VALUE: This is scrap value. It is the amount which can be recovered when the assets is sold at the end of useful life.

ESTIMATED USEFUL LIFE: This is the expected number of years through which an asset can last.

STRAIGHT LINE METHOD (SLM)

$\Rightarrow$ Cost-Est. Value
          Years of useful life

EXAMPLE 3.32: The cost of machine is ₦10,000. The estimated/residual value is ₦4,000. It is expected to last for 4yrs, find the depreciation.

SOLUTION

1. $\dfrac{10,000 - 4,000}{4} = \dfrac{6000}{4}$

      = **₦1,500** in 4 years

Year	Cost of Asset	Depreciation Rate	Accumulated Depreciation	cost of asset less accumulated Net-Book Role)
1	10,000	1500	0 +1500=1500	10,000-1500=8500
2	10,000	1500	1500+1500=3000	10,000-3000=7000
3	10,000	1500	3000+1500=4500	10,000-4500=5,500
4	10,000	1500	4500+1500=6000	10,000-6000=4000

Considering the ways of depreciation and knowing that larger amount of depreciation is deemed to occur in the early years and smaller amount in the later years,

Depreciation is computed using the formula,

$$D_{rate} = 1 - \sqrt[N]{\frac{\text{Estimated or residual value}}{\text{Cost}}}$$

EXAMPLE 3.33:

A motor cost ₦6,400 in 2001, it will be kept for 5yrs and then sold at an estimated/residual value of ₦200. Compute the depreciation rate.

SOLUTION

$$D_{rate} = 1 - \sqrt[5]{\frac{200}{6,400}}$$

$$= 1 - \sqrt[5]{1/32}$$

$$= 1 - \sqrt[5]{1/2^5}$$

$$= 1 - \sqrt[5]{2^{-5}}$$

$$= 1 - (2^{-5})^{1/5}$$

$$= 1 - 2^{-1}$$

$$= 1 - \frac{1}{2}$$

$$= \frac{1}{2} = 0.5$$

$$\therefore 0.5 \times 100 = \mathbf{50\%}$$

YEARS	COST OF MOTOR	SOLUTIONS
So, 1st yr.	Depr.(50% of 6400) 3200 ⟶	6400 -3200 3200
2nd yr.	Depr. (50% of 3200) 1600 ⟶	3, 200 - 1, 600 1,600
3rd yr.	Depr. (50% of 1600) = 800 : . ⟶	1600 - 800 800
4th yr.	Depr (50% of 800) = 400 ⟶	800 - 400 400
5th yrs	Depr(50% of 400) = 200 ⟶	400 - 200 200

3.10 RATIOS AND PROPORTIONS

RATIOS

These are fractions which express variations in the data irrespective of actual absolute size of the data. This is a fraction that compares two quantities that are measured in the units. One quantity is the numerator of the fraction, and the other is the denominator.

EXAMPLE 3.34: If there are 3 boys and 24 girls in a class, we say that the ratio of the number of boys to the number of girls in that class is 3:24 or 1:6 in a simplified manner.

EXAMPLE 3.35: If 40% of the students in a Secondary School are females, what is the ratio of female students to male students?

SOLUTION

Note: Problems involving present implies that the number 100 should be used.

∴ Assuming there exist 100 students, then 40 of them are females and

(100-40) are males meaning (*100-40*) =60

∴ The ratio of females to males is

$$\frac{40}{60} = 46:60$$

$$= 4:6$$

$$= \mathbf{2:3}$$

EXAMPLE 3.36: The measure of the ratio of acute angles in a right angled triangle is 5:25. What is the measure of the largest acute angle?

SOLUTION

From the ratio, let the measure of the smaller angle be 5x

And, let the measure of the larger angle be 25x

Since the sum of the measures of the two acute angles of a right angled triangle is 90^0.

So, $5x + 25x = 90$

$30x = 90^0$

$\therefore$ The measure of the larger angle is $25 \times 3 = 75^0$ and the measure of the smaller angle in $5 \times 3 = \mathbf{15^0}$

However ratio can be extended to three or four or more terms.

EXAMPLE 3.37: If the ratio of FANTA to COKE to ORANGE drinks sold at Mr. Biggs supermarket Gusau was 6:5:3 on a particular day. What percentage of the COKE drink was sold?

SOLUTION

Total ratio of the drinks is $6+5+3 = 14$

$\therefore$COKE drink will be made up of $= 5$

$6+5+3 = {}^5/{}_{14}$

$= 0.3571 \times 100$

$\mathbf{35.7\%}$

$\therefore$35.7% out of the total drinks was COKE drink.

PERCENTAGES

These are ratios expressed with 100 as the denominator, although this is usually not written.

EXAMPLE 3.38: If the total expenditure is ₦30,000 out of which ₦12,000 was spent on food, then the ratio of food to total expenditure is:

SOLUTION

$$\frac{₦12,000}{₦30,000} = {}^{2/5}$$

However, the percentage of food is $\dfrac{12,000}{1} \div \dfrac{30,000}{100}$

$$= \frac{12000}{1} \times \frac{100}{30,000} = \frac{120}{3} = \mathbf{40\%}$$

PROPORTION

This is an equation which states that two ratios are equivalent, since ratios are just fraction.

Any equation such as $^3/_6 = {}^{15}/_{30}$ in which each side is a simple fraction is a proportion.

Algebraically, $a/b = c/d$ is a proportion meaning, cross multiplying, $ad = bc$.

EXAMPLE 3.39: If 5 workers can paint a certain number of houses in 24 days. How many days will 15 workers take if they work at the same rate to do the same job?

SOLUTION

Note: The more worker the less the crime.

So, if 5 workers = 25 days

∴ 15 workers will be x days

Cross multiplying,

$$5x = 24 \times 15$$

$$x = 360/5 = \textbf{72}$$

∴ It will take 72 days.

RATE

This is a fraction that compares two quantities that are measured in different units.

Units are time expressed in terms of minute, hour, day etc of work he or she can do in a unit time.

EXAMPLE 3.40: Mr. Okoko read 12 pages of the New Weekly Legacy Newspapers in 5 minutes. At this rate, how many pages can he read in 45 minutes?

SOLUTION

Set up a proportional statement and cross multiply.

If Mr. Okoko can read 12 pages for 5 minutes,

∴ He can read x pages in 45 minutes

Meaning 12 pages = 5min

x pages = 45min

$5x = 12 \times 45$

$5x = 540$

$x = 540/5 = \mathbf{108}$

So, he can read 108 pages in 45 minutes.

EXAMPLE 3.41:

If a worker can do a job in 6 days, then he can do $1/6$ of the job in 1 day.

Given the various time in which each of a number of people can complete a job, given various times in which each of a number of people can complete a job and finding the time it will take to do the job if all work together is.

1ST STEP:

Invert the time of each to find how much each can do in unit time.

2ND STEP:

Add these reciprocals to find what part of the job all working together can do in unit time.

3RD STEP:

Invert this sum to find the time it will take all of them together to do the whole job.

EXAMPLE 3.42:

If it takes Mr. A two (2) days to dig a certain pit and Mr. B can digit in 4days, and Mr. C. in 8days, how long would it take all of them to do the job?

SOLUTION

Mr. A can do it in   2 days

i.e He can do it in $1/2$ days

Mr.B can do it in   4days

i.e He can do it in   ¼ days

Mr. C can do it in   8days

i.e. He can it in   1/8 days

So, $\frac{1}{2} + \frac{1}{4} + \frac{1}{8} = \frac{7}{8}$

Mr. A, B and C can do it in $\frac{7}{8}$

∴. It will take them $\frac{8}{7}$ days or $1^{1/7}$ days to complete the job (the in verse of the reciprocals of the individual days or work)

NOTE. The reciprocal of the work done in unit time is the time it will take to do the **COMPLETE JOB.**

REVISION EXERCISE 3.00: A machine was bought in 1990 for **₦5,500**. It was estimated to be **₦500** after 4 years of use. Then calculate the depreciation.

Chapter Four

VARIATIONS

Here, we ought to identify the different types of variations viz: Direct, Inverse, Joint and Partial

4.01 <u>DIRECT VARIATION</u>

A quantity is said to vary directly as another, x that is: y ∞ x or read as y is directly proportional to x. If ratio of y to x is a constant

Thus: $y/x = y_1/x_1 = y_2/x_2 = y_3/x_3 \ldots y_n/x_n$ where $X_n \neq 0$

EXAMPLE 4.00: The Volume, V of a sphere varies (directly) as the cube root of its radius. If the Volume of the sphere of radius 2cm is 4.188cm³, find {correct to three significant figures}, the Volume of a sphere of radius 4cm.

SOLUTION

$V \infty r^3$

$V = Kr^3$ .... equation. i

$K = V/r^3$ .... equation. ii

$K = \dfrac{4.188}{(2)^3} = \dfrac{4.188}{8}$

= 0.5235

To find V again when r = 4

V   = 0.5235 x (4)³

         = 0.5235 x 64

         = 33.504

V   = **33.5cm³**

4.02 INVERSE VARIATION

y is said to vary inversely as x or read as y is inversely proportional to x.

If y varies inversely as x, it implies that, y ∞ 1/x

Introducing the constant of proportionality,{ k }

    y = k(1/x)

    y = k/x ; k = yx

Thus x, y = x^2y^2 = x^3y^3 = x^ny^n = k

EXAMPLE 4.01: If r varies inversely as the square root of h and r = 6

when h = 4. Find r when h = 9 and find h when r = 5

SOLUTION

$$r \propto 1/\sqrt{h}$$

$$\therefore \quad r = k(1/\sqrt{h})$$

$$r = k/\sqrt{h}$$

so, $\quad k = r\sqrt{h} = 6\sqrt{4}$

$$= 6 \times 2 \text{ So, } k = 12$$

To find r

$$r = {}^{k}/_{\sqrt{h}} = {}^{12}/_{\sqrt{9}} = {}^{12}/_{3} = 4$$

If $\quad k = r\sqrt{h}$

$$k^2 = r^2 h$$

Reversing

$$r^2 h = k^2$$

$$h = k^2/r^2 \text{ If } k = 12, r = 5$$

$$h = (12)^2/(5)^2 = 144/25 = \mathbf{5.76}$$

4.03 <u>JOINT VARIATION</u>

This will involve relationship between two or more variables. It can be for either direct, inverse or both.

EXAMPLE 4.02:

$y \propto 1/x^2$ and $y \propto z^2$

$\therefore$ $y \propto z^2/x^2$, $y = K(z^2/x^2)$

$\Longrightarrow$ $y = kz^2/x^2$

$Kz^2 = yx^2$

$\therefore$ yx^2/ z^2 if $y = 12$, $x = 2$ and $z = 4$

$\Longrightarrow$ $K = 12(2)^2/(4)^2 = (12 \times 4)/16$

$K = 3$

Finding y when $z = 6$ and $x = 4$

$\Longrightarrow$ $yx^2 = kz^2$

$y \;\; = Kz^2/x^2$

    $= 3(6)^2/(4)^2$

    $= (3 \times 36)/16$

$y \; = \mathbf{27/4.}$

4.05 <u>PARTIAL VARIATION</u>

See ***Simplified Mathematics Series II for Secondary Schools with Answers.***

REVISION EXERCICES 4.00:

1. The thickness of a book varies directly with the number of pages in the book. A book is 1.8cm thick and contains 350 pages. What is the thickness of the first 84 pages of the book?

2. If r is inversely proportional to h, and r =5 when h = 12.

   {a} find r when h = 20 { b} find h when r = 45.

3. x varies jointly as y and z when y=9 and z = 2

   { a} find the relation between x, y and z {b} find x when y = 14 and z= 12

Chapter Five

ALGEBRAIC PROCESSES

5.01 <u>SIMPLIFICATION OF ALGEBRAIC EXPRESSION</u>

Algebraic expression comprises of numerical and variable combined with any basic arithmetic signs i.e. +, x,-, ÷, etc

Algebraic expression connected by symbols of plus { + } and minus {-} sign is known as **ALGEBRAIC SUM** and the connected parts are called its **FACTORS**.

EXAMPLE 5.00:

$8x - 4xy + 6y - 3$ is an algebraic sum of four terms 8x, 4xy, 6y and 3.

The factors of 4xy are : **4xy, 4, x** and **y**. The numerical value 4 is known as **CONSTANT** or **COEFFICIENTS**.

Coefficient of a term in an expression is the number which is multiplied by one or more variables or power of variables in that term. It can be positive or negative. In the case of polynomials $\{-3x^3 + 4 x^2 x^3 + 1 = 0 \}$, you have to add or subtract the coefficients in the like terms.

EXAMPLE 5.01: Given the polynomial,$-3x^3 + 4x^2 x^3 + 1 = 0$

Bringing like terms together,

$$-3x^3 x^3 + 4x^2 + 1 = 0$$

$$-4x^3 + 4x^2 + 1 = 0$$

So, the coefficient of x^3 = **-4**, x^2 = **+4** , constant = 1.

EXAMPLE 5.02: $P^3 + 2P^2 - 2P + 5P - 1,$

{ Bringing like terms together }

$$P^3 + 2P^2 + 3P - 1$$

So, the coefficient of P is **+3.**

EXAMPLE 5.03: 5x-3, the coefficient of x is **5** and -3 is the constant.

EXAMPLE 5.04: $3xy^2 - 7z$, the coefficient of z is **-7** and that of xy^2 is **+3.**

EXAMPLE 5.05: $P^3 + 2P^2 - 5P - 1$, coefficient of P^3 is **+1**, for P^2 is **2** and P is **-5**.

A product of more than one identical factor is depicted as an index or exponent and the factor is called the base of the index.

EXAMPLE 5.06: Given 2 x 2 x 2 x 2 = 2^4 = **16**

Shows that 2 is the base of the index,

And, 4 is the index or power {exponent} because it indicates the power to which the base is raised. The algebraic terms having no difference rather than coefficient are called **LIKE TERMS** otherwise **UNLIKE TERMS**.

EXAMPLE 5.07: Like terms: mn^2, $6mn^2$, $-2mn^2$

Unlike terms: **6ax, 3yb, 6jk**

NOTE: When solving algebraic expression, we group like terms together

EXAMPLE 5.08: Simplify the following 21m + 7n - 15m + 12n

SOLUTION 1st Step rearranging the terms together, we have

21m - 15m + 7n + 12n

= 6m + 19n

5.02 <u>BRACKETS</u>

These show the manner in which operation must be done. When using bracket, terms inside it are taken as whole.

EXAMPLE 5.09: 4 + (3 - 2) means that 2 must be subtracted from 3 and the difference is added to 4.

Therefore 4 + (1) = **5**

Also, 2 + (3 + 1) means that 1 must be added to 3 and the sum added to 2 to give **6**

Generally,   a + (b + c) = a + b l c

a + (b - c) = a + b - c

Also $\qquad$ a - (b + c) = a - b - c

$\qquad$ a - (b - c) =-b + c + a = c + a - b

Thus, 14 - (8 - 12) =-8 + 12 + 14 if a = 14, b = 8 and c = 12 and

$\qquad$ b < c = 4 + 14 = 18

$\qquad$ and/or c + a - b = 12 + 14 - 18

$\qquad$ = 26 - 8 = **18**

EXAMPLE 5.10:

Simplify $\qquad$ 3(x - 2) - 2(x + 1)

Multiplying out

$\qquad$ = 3x - 6 - 2x - 2

$\qquad$ = 3x - 2x - 6 - 2

$\qquad$ = **x - 8**

5.03 <u>FACTORIZATION</u>

Writing the expression as a product of two or more algebraic expression

EXAMPLE 5.11: Factorize: $15a^2 - 3$

$\Longrightarrow$ **$3(5a^2 - 1)$**

EXAMPLE 5.12: Factorize: $25x^2+5$

$\Longrightarrow$ **$5\{ 5x^2+1\}$**

EXAMPLE 5.13: Factorize: $3a^2-9a$

$\Longrightarrow$ **$3a \{a-3\}$**

EQUATIONS INVOLVING FOUR TERM

EXAMPLE 5.14: Factorize: $9a^2 - 6ab + 6ab - 4b^2$

$= 3a (3a - 2b) + 2b (3a - 2b)$

EXAMPLE 5.15: Factorize: $a^2x - b^2y + b^2x + a^2y$

SOLUTION: The above is equal to

$(a^2x + a^2y) (-b^2y - bx^2)$

$a^2 (x + y) -b^2(y + x)$,the common factor among the two terms are x+y or y+x

d/4 $\{ a^2+b^2\} \{ x+y \}$ or $\{ a^2 b^2\} \{ y+x \}$ are the products.

$\Longrightarrow$ $(a^2 - b^2) (x + y)$

Note: $(a^2 - b^2) = (a - b) (a + b)$

Also $(a - b)^2 = a^2 - 2ab + b^2$

$\therefore (a^2 - b^2)(x + y) = (a - b)(a + b)(x + y)$

5.04 FACTORIZATION OF QUADRATIC EQUATIONS

EXAMPLE 5.16:

However in solving quadratic equation, an expression like

$35 - 2b + b^2$ is written as; $35 - 2b - b^2 = 0$

$\Longrightarrow$ $-b^2 - 2b + 35 = 0$

Multiplying both sides by (-1), becomes

$b^2 + 2b - 35 = 0$, where a = +1, b = +2 ,c =-35

Let   b = x

$\therefore$ $x^2 + 2x - 35 = 0$

Since factors of 35 are 5 and 7, 5 - 7 = 2

We, then have

$x^2 + 7x - 5x - 35 = 0$ equation. (1)

$x^2 + 2x - 35 = 0$

Using equation (1)

$(x^2 + 7x)(-5x - 35) = x(x + 7) - 5(x + 7)$

$\therefore$ **x - 55 = 0 or x + 7 = 0**

AC TEST IN FACTORIZATION

1. $x^2-3x-18$

a $=1$,b $=-3$ and c $=-18$,then we look for two numbers m and n such that their product **ac** ,and their sum is b.

In this case, mn $=-18$ and m+n $=-3$

mn	m+n
$1\{-18\} =-18$	$1+\{-18\} =-17$
$2\{-9\} =-18$	$2+\{-9\} =-7$
$3\{-6\} =-18$	$3+\{-6\} =-3$
$6\{-3\} =-18$	
$9\{-2\} =-18$	
$18\{-1\} =-18$	

Since,3 and-6 are the two integers with a product of **ac** and a sum of b we then say that m=3 and n =-6

So, $x^2-3x-18$ is factorable if m and n are the values.

2. $x^2-24x + 23$

a $=1$,b $=-24$ and c $= 23$

ac $= 1\{ 23\} = 23$

b$=-24$

So, **m = 23, n + m =-24**, we calculate the integer pairs as;

mn	m+n
1{ 23} = 23	1+23 = 24
-1{-23} = 23	-1+{-23} =-24

m =-1, m =-23

So, x^2-24x + 23 is factorable.

3. x^2-11x + 8

   a = 1, b =-11 and c = 8

So, ac = 8 and b =-11, thus mn = 8 and n + m =-11.

We calculate the integer pairs as ;

mn	m+n
1{ 8} = 8	1 + 8 = 9
2{ 4} = 8	2 + 4=6
-1{-8} = 8	-1+{-8} =-9
-2{-4} = 8	-2+{-4} =-6

There are no other pairs of integers with a product of 8, and non of these pairs has a sum of-11.

So, **the trinomial x is not factorable.**

REVISION EXERCISE 5.00:

1. Factorize the following quadratic equations

i. $u^2 + 18u + 72$

ii. $12x^2 + 14x - 20$

iii. $16x^2-9$

iv. $25y^2 - 4$

2. Using AC Method , solve $2x^2+7x-15$.

5.06 <u>PERFECT SQUARES AND ALGEBRAIC FRACTIONS</u>

PERFECT SQUARES

This is an expression having a number or a polynomial which is second power of a number or polynomial.

Having $\quad(x + y)^2 = x^2 + 2xy +y^2$ $\qquad$ equation. (1)

And $\quad(x - y)^2 = x^2 - 2xy + y^2$ $\qquad$ equation. (2)

We have difference of two squares as

$\quad x^2 - y^2 = (x + y) (x - y)$ $\qquad$ equation .{ 3 }

Also $-(a - b) =-a + b = b-a$

Showing that the negative of a - b = b - a, we can factorize the basic operations of arithmetic.

EXAMPLE 5.17: Evaluate $(98)^2$

SOLUTION

This is equal to $(100 - 2)^2$

Using equation (2) above, where x = 100 and y = 2

∴ $(100)^2 - 2(100)(2) + (2)^2$

     = 10000 - 400 + 4

     = 10000 - 396 = **9604**

EXAMPLE 5.18:

Evaluate $(3.2)^2 - (1.8)^2$

SOLUTION

Using equation { 3 } , where x = 3.2, y = 1.8

∴   (3.2 + 1.8) (3.2 - 1.8) = (5)(1.4) = 7

EXAMPLE 5.19: Multiply 112 by 88

SOLUTION: { 100-12 } { 100 + 12 }

             100 { 100-12 } +12 { 100-12 }

             100000 - 1200 +1200-144

$$10000 - 144 = \mathbf{9856}$$

REVISION EXERCISE 5.01:

   i. Multiply 52 by 48

   ii. Multiply 77 by 83

   iii. Multiply 66 by 54

   iv. Evaluate $\{ 200 - 2 \}^2$

SOLUTION

   i. $\{ 50 + 2 \} \{ 50 - 2 \}$

      $50 \{ 50 - 2 \} + 2 \{ 50 - 2 \}$

      $2500 - 100 + 100 - 4$

      $= 2500 - 4$

      $= \mathbf{2496}$

   ii. $\{ 80 - 3 \} \{ 80 + 3 \}$

      $80 \{ 80 + 3 \} - 3 \{ 80 + 3 \}$

      $6400 + 240 - 240 - 9$

      $6400 - 9 = \mathbf{6391}$

   iii. $\{ 60 + 6 \} \{ 60 - 6 \}$

60 { 60-6 } + 6{ 60 - 6}

3600 - 360 + 360-36

3600-36 = **3564**

iv. Using equation { 2 } where x = 200 and y = 2

d/4 { 200 } - 2{ 200} { 2 } +{ 2}²

40000 - 800 + 4 = **39204**.

ALGEBRAIC FRACTIONS

Here, we encounter the problems of undefined expression. That is when the denominator of an expression is equal to zero.

EXAMPLE 5.20: For what value of y is the expression

$$\frac{y + 2}{y^2 - 3y-10} \qquad \text{undefined ?}$$

SOLUTION

The expression is undefined when $y^2 - 3y - 10 = 0$

$\Longrightarrow$ (y + 2) and (y - 5) = 0

y + 2 = 0 ; **y =-2** or **y** - 5 = 0 ; **y = 5**

Substituting the value of y for 5,

$$\frac{1}{y-5} = \frac{1}{5-5} = \frac{1}{0}$$

EXAMPLE 5.21: For what value of x is the expression

$$\frac{1 - x^2}{(x + 1)^2}$$ undefined ?

SOLUTION

The expression is undefined when $\{x + 1\}^2 = 0$

$\implies \{x + 1\}^2 = 0$

$x + 1 = 0$ or **x = -1**

EXAMPLE 5.22:

The expression $\dfrac{1 - x^2}{x - x^2}$ When $x \neq 0$ is undefined when

$x\{1-x\} = 0$, that is to say $x = 0$ or $1 - x = 0$

So, **x = 0** or **x = 1**.

5.07 <u>SUBSTITUTION PROCESSES</u>

EXAMPLE 5.23: For what value of x is the expression undefined ?

$$\frac{xy^2 - x^2y}{x^2 - xy} \qquad \text{when x} = -2, \text{y} = 3$$

$$\frac{(-2)(3)^2 - (-2)^2(3)}{(-2)^2 - (-2)(3)}$$

$$= \frac{(-2 \times 9) - (4 \times 3)}{(4-6)}$$

$$= \frac{-18 - 12}{10} = \frac{-30}{10} = -3$$

Chapter Six

SIMPLE EQUATIONS

What is an equation? This is a statement of equality. If the equality is true only for certain values of the unknown qualities involved in the equation ,then it is sometimes called *conditional equation* and the equality sign { = } is used ;but if the equality is true for all values of the unknown quantities involved ,the equation is called *identity* and the equivalence { ≡ } sign is used. **NB:** The later is treated in **Simplified Mathematics Series II for Secondary Schools with Answers.**

6.01 CONDITIONAL EQUATION

EXAMPLE 6.00:

Given a simple equation: 2x - 9 = 15, find the value of the unknown?

SOLUTION:

$$2x - 9 = 15$$

Bringing like terms together,

$$2x = 15 + 9$$

$$2x = 24$$

So, $x = {}^{24/2} = \mathbf{12.}$

EXAMPLE 6.01: Find the value of x in the following: 3x-12=0

$$x = {}^{12/3}$$

$$x = \mathbf{4}$$

EXAMPLE 6.02: Find the value of x in the following: 4x-1=1

Bringing like terms together,

$$4x=1+1$$

$$4x = 2$$

$$x = 2/4 = {}^{1/2}$$

REVISION EXERCISE 6.00:

1. x - 3x = 7

2. 6x-x + 2 = 0

3. x-4x + 4 =-5

4. 2-3x + 1 = 0

Chapter Seven

SIMULTANEOUS EQUATION

Two or more equations that are true at one and at the same time, and are therefore satisfied by the same values of the unknowns involved are called simultaneous equations.

For instance, $2x + y = 3$

$3x + 4y = 2$ have solutions $\{\, x = 2 \text{ and } y = -1\}$

METHODS

Methods of solution include:

1) Substitution.

2) Elimination.

3) Graphical method.

EXAMPLE 7.00: Using the three methods, Solve

$2x - y = -1$

$2x + y = 9$

7.01 <u>SUBSTITUTION METHOD</u>

$2x - y = -1$ ....equation (i)

$2x + y = 9$ ....equation (ii)

$2x = -1 + y$

$x = -\dfrac{1 - y}{2}$ ....equation (iii)

Substitution equation (iii) into equation (ii), becomes

$2 \left[\dfrac{-1 + y}{2}\right] + y = 9$

$-1 + y + y = 9$

$2y = 10$

$y = \mathbf{5}$

To find x,

We substitute this value back to equation (ii)

$2x + 5 = 9$

$2x = 9 - 5$

∴ $\quad x = {}^{4}/_{2} = \mathbf{2}$

x = 2, y = 5

7.02 <u>ELIMINATION METHOD</u>

$2x - y = -1$ ....equation (i)

2x + y = 9 ....equation (ii)

Adding equation (i) and (ii)

4x =8

$\therefore$ $= {}^8/_4$

So   **x = 2**

Putting x into equation (i)

2(2) - y =-1

4 - y =-1

Bringing like term together

4 + 1 = y

$\therefore$ **y = 5**

2x - y = 1, 2x - 1 = y

2x + y = 9, 2x - 9 =-y or y = 9 - 2x

Forming a table

X	0	1	2
y = 2x - 1	-1	1	3
y = 9 - 2x	9	7	5

EXAMPLE 7.02:

Solve the simultaneous equation :

3x-2y = 7   equation i { *multiply by 1* }

x + 2y =-3   equation ii { *multiply by 3*}

3x-2y = 7 ....equation (iii)

3x + 6y =-9 ....equation (iv)

{ iii-iv }   -8y = 16

          y = 16/-8 =-2,

To find the value of x, we substitute the value of y into equation ii,

x+2y =-3

x + 2 {-2 } =-3

x +-4 =-3

x - 4 =-3

x =-3 + 4

x = 1

y =-2 ; x = 1 { **WAEC,JUNE 2011**}

REVISION EXERCISE 7.00:

Solve the following simultaneous equation

1. 3x + 5y = 4

    4x + 3y = 5   { *JAMB 1982* }

2. x + y = 3/2

$x - y = 5/2$ { **SSCE 1995**}

3. $x + y = 2$

$3x-2y = 1$ { **SSCE 2007**}

$3x-2y = 21$

$4y + 5x = 5$ { **SSCE 1999** }

7.03 <u>GRAPHICAL METHOD:</u> See *Simplified Mathematics Series II for Secondary Schools with Answers.*

Chapter Eight

SOLUTIONS OF QUADRATIC EQUATIONS AND WORD PROBLEMS LEADING TO LINEAR EQUATIONS.

This involves the use of word in forming equations, and letters are used to represent quantities in solving word problems.

8.01 **METHODS OF SOLUTION**

(i) Choosing letters to represent the number required

(ii) Translating each statement given in the equation to a statement containing the chosen letters.

(iii) Formulating equation by linking up the parts of the equation

EXAMPLE 8.00:

A man is 3 times as old as his daughter; 8 years ago the product of their ages was 112. Calculate their present age.

SOLUTION

Let the man's age be = x

Let the daughter's age be = y

$\therefore$ $x = 3y$...equation (i)

8 years ago, their ages are

    $x - 8 = 3(y - 8)$

from the statement, the product of their ages 8 years ago

    $= 112$

: $-(x-8)(y-8) = 112$

Factorizing

    $x(y-8)-8(y-8)$

    $xy - 8x - 8y + 64 = 112$

    $xy - 8x - 8y + 64 = 112$

    $xy - 8x - 8y = 112 - 64$

    $xy - 8x - 8y = 48$

    $(3y)y - 8(3y) - 8y = 48$

    $3y^2 - 24y - 8y = 48$

Using formula method of quadratic equation,

$$y = \frac{-b \pm \sqrt{b^2 - 4ac}}{2a} = \frac{-32 + \sqrt{(-32)^2 - 4(3)(-48)}}{2 \times 3}$$

$$=\frac{32+\sqrt{1024-576}}{6} = \frac{32+\sqrt{1600}}{6}$$

$$=\frac{32+40}{6}$$

$$y_1 = \frac{32 + 40}{6} \text{ or } y_2 = \frac{32 - 40}{6}$$

$$y_1 = 12 \text{ or } y_2 = -1.33$$

∴   **y = 12** {when *selecting the positive part*}

So, applying back to equation (i)

$$x = 3(12)$$

x = 36 years.

EXAMPLE 8.01: The sides of a triangle are (x+4)cm, x cm and (x-4)cm respectively. If the cosine of the largest angle is $^1/_5$, find the value of x.

SOLUTION

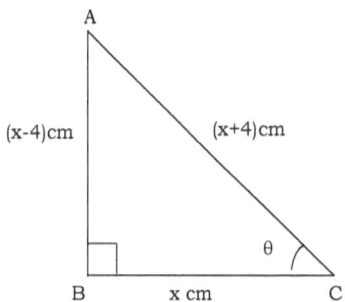

Using **SOHCAHTOA**

Cosine θ = $\dfrac{Adj}{Hypo}$

Cos (x+4)= $\left(\dfrac{x-4}{x+4} = \dfrac{1}{5}\right)$

Cross multiplying in the bracket

$1(x+4) = 5(x-4)$

$x+4 = 5x-20$

$x-5x = -20-4$

$-4x = -24$

$x = \dfrac{24}{4} = 6$

∴   **x = 6**

EXAMPLE 8.02:

A mother is three times old as her son. In 12 years time she will be twice as old as her son. Find the age of the mother.

SOLUTION

Let mother's age be x

Let son's age be y

: .   x = 3y.. equation 1

In 12 years time

$$x + 12 = 2(y + 12)$$

$$x + 12 = 2y + 24$$

$$x = 2y + 24 + 12$$

$$x = 2y + 12$$

But $x = 3y$ from equation 1

So, $3y = 2y - 12$

Bringing like terms together.

$$3y - 2y = 12$$

∴ $y = 12$

So, the mother will be,

$$x = 3y$$

$$= 3(12)$$

$$= \textbf{36years}$$

REVISION EXERCISES 8.00:

1) Think of a number, subtract 12 from its square the result is 30 added to number, and find the number.

2] Seven years ago the age of a father was three times that of the son, but in six years time the age of the son will be half the father.

Representing the present age of the father and don by x and y respectively

The two equations relating x and y are:

{*JAMB 1982*}

{3} Fathers age = x ,Sons age = y , If a father is 5 times as old as his son.

It means that x = 5y

If x + y = 60 , How old is the father?

Having ,   5y+ y = 60

              6y = 60

              y = 60 / 6 = 10

So,   y = 10

And fathers age , x = 5y

              = 5{ 10 }

              = **50 years**

Son's age, y = 10 years { **SSCE 2011**}

Chapter Nine

CHANGE OF SUBJECT OF THE FORMULA

9.01 CHANGE OF SUBJECT OF THE FORMULA

EXAMPLE 9.00: Make h the subject of the formula if given

$$S = 2\pi rh + 2\pi r^2$$

SOLUTION

$$S = 2\pi rh + 2\pi r^2$$

$= 2\pi r(h + r)$,reversing the above equation

$(h + r)\, 2\pi r = S$

$$h + r = \frac{S}{2\pi r}$$

$$\mathbf{h = \frac{S-r}{2\pi r}}$$

EXAMPLE 9.01: Make T the subject of the formula

Given $\qquad \dfrac{ar}{1-r} = \sqrt[3]{\dfrac{2r+T}{a+2T}}$

SOLUTION

$$\left[\frac{ar}{1-r}\right]^3 = \frac{2r + T}{a + 2T}$$

$$\frac{a^3 r^3}{(1-r)^3} = \frac{2r + T}{a + 2T}$$

$a^3 r^3(a + 2T) = 2r + T(1 - r)^3$,means that $a^4 r^3 + 2Ta^3 r^3 = 2r(1 - r)^3 + T(1 - r)^3$

$a^4r^3 - 2r(1-r)^3 = T(1-r)^3 - 2a^3r^3$

$\therefore$ **$T = \dfrac{a^4r^3 - 2r(1-r)^3}{(1-r)^3 - 2a^3r^3}$**

EXAMPLE 9.02:

Make a the subject if $a/x + b/y = a/y$

SOLUTION

Finding the Lowest Common Factor,

$$\frac{ay^2 + xyb = xya}{xyy}$$

Having , $\qquad ay^2 + xyb = xya \qquad$ from the numerator,

$\qquad\qquad\qquad xyb = xya - ay^2$

$\qquad\qquad\qquad xyb = a\{ xy - y^2\}$

a = xyb

xy-y²

= xyb

y{ x-y}

a = xb/x-y   { **NECO 2011**}

REVISION EXERCISE 9.00:

1) Given $\frac{P}{\sqrt{2}} = \sqrt{\frac{r}{r + q}}$, make r the subject

2) Make x the subject given

$$\frac{1 - ax}{1 - ax} = \frac{P}{q} i$$

3) Make c the subject, if

$$r = 1 - ^a/_5 \left[b^{-3c}/_7 \right]$$

Chapter Ten

LINEAR INEQUALITIES

This has it form as ax + b > 0

where a and b are constants. Having one variable as:

"x" and ">" (Greater than).

Others are:

<; Less than

≥; Greater than or equal to

≤ ; Less than or equal to

In problem solving, we should note that

inequality sign is:

1. PRESERVED IF

a) The same quality or number is added to or subtracted from both sides of the inequality.

b. Both sides of inequality are multiplied by or divided by the same positive number.

2. REVERSED IF

Both side of the inequality (ics) are multiplied or divided by the same negative number

EXAMPLE 10.00: Find the range of values of x such that 2(3x-2) ≥ x +5

SOLUTION:

2(3x-2) ≥ x +5

6x-4 ≥ x +5

Bringing like terms together,

6x-x ≥ 5 + 4

5x ≥ 9

∴   **≥ 9/5**

EXAMPLE 10.01:

Represent each of the following inequalities on a number line

(i) 3x + 2 ≤8

(ii) 2-3x>6

{iii}-1<3x+5<14

SOLUTION:

(i) 3x +2≤8

Bringing like term together

$3x \leq 8\text{-}2$

$3x \leq 6$

$x \leq 6/3$

∴ $x \leq \mathbf{2}$

10.01: NUMBER LINES:

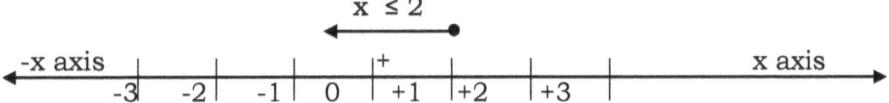

Description of the number line;

1. We use the x-axes because, we are solving for x

2. The arrow started at 2 as a value of x

3. The arrow pointed towards the inequality sign

4. The point of starting (●) was shaded to confirm

That types inequality.

$2\text{-}3x > 6$

$-3x > 6\text{-}2$

$-3x > 4$

$x < \text{-}4/3$

: .   **x <-1.33**

NUMBER LINE:

x<-1.33

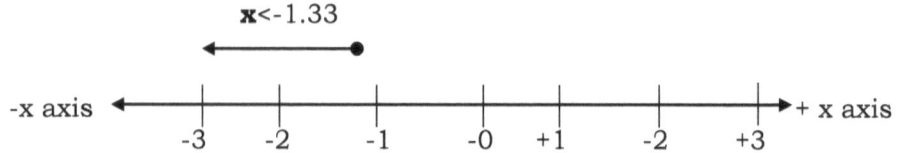

{iii}splitting the equation ,it becomes

-1 < 3x + 5 and 3x +5 < 14

Solving for the first part,-1-5 < 3x

-6 < 3x

-6/3 = x

-2 = x

and   3x + 5 < 14

3x < 14-5

3x < 9

x < 3

So,   -2 < x < 3

REVISION EXERCISE 10.00:

1. Solve for k , if $2k-2 \leq \dfrac{k+5}{2}$

10.02 PRACTICAL APPLICATIONS OF LINEAR INEQUALITIES IN TWO

VARIABLES.

Given ax + by c ≥ 0 as the equation of a straight line dividing the plane into two so that one part satisfies the inequality ax + by + c < 0, and the other satisfies the inequality ax + by + c > 0

Graphically

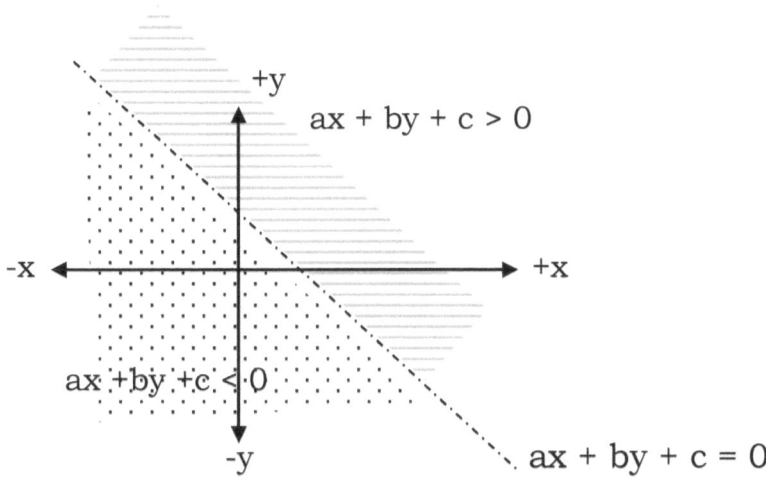

As seen above, the dotted line in ax + by-c = 0, showing that it is not part of the inequality. For ≤ or ≥ means, that ax + by + c = 0 will be drawn "thick" as a continuous line in which ax + by + c = 0, will be part of the inequality as shown below.

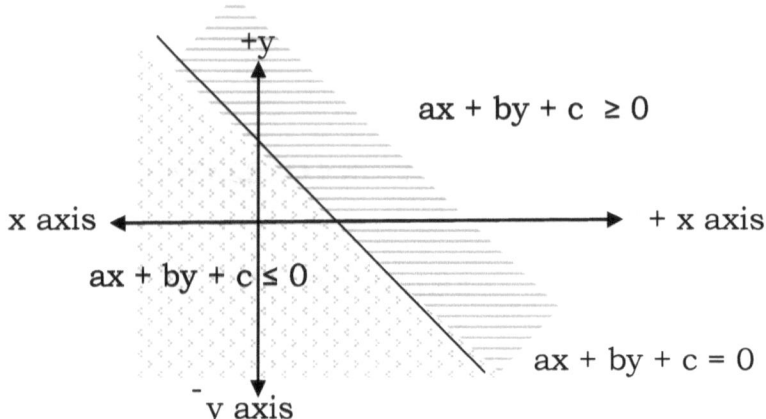

Chapter Eleven

CONGRUENCY AND SIMILAR PROPERTIES OF TRIANGLES

11.01 **TRIANGLE**

According to the *Oxford Learners Dictionary,* is defined as a flat shape with three straight sides and three angles in Geometry having a total of 180^0.

Jaggi (2006) defined it as a plane figure bounded by three sides, or a polygon with three sides.

Congruency between two triangles exist if and only if any of the following conditions are satisfied

1} **SSS**

Having

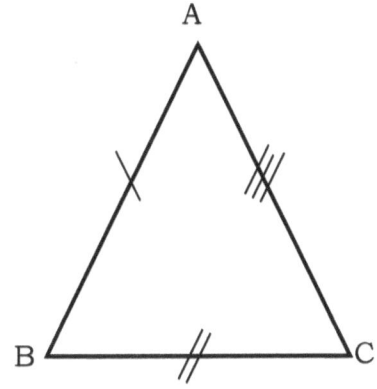 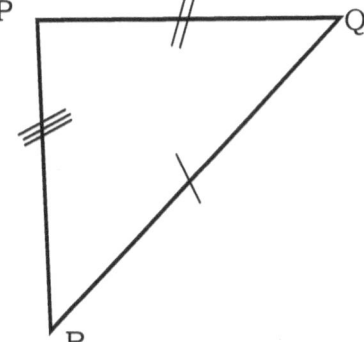

$\overline{AC} = \overline{PR}$ (equal sides **S**)

$\overline{BC} = \overline{PQ}$ (equal sides **S**)

$\overline{AB} = \overline{RQ}$-(equal sides **S**)

∴ Congruency exist between △ ABC and △PQR

2}. SAS

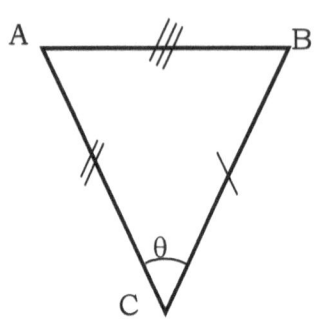 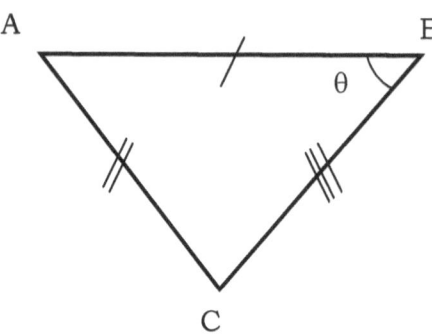

$\overline{AB} = \overline{RQ}$ (equal sides **S**)

$A\hat{C}B = P\hat{Q}R$ (equal angle **A**)

$CB = PR$ (equal side **S**)

∴ Congruency exist between △A$\hat{B}$C and △P$\hat{Q}$R

3}. AAS

Having

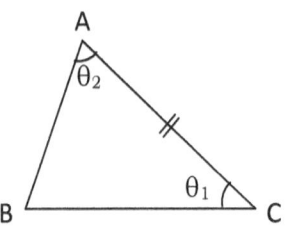 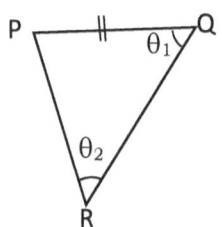

$B\hat{A}C = P\hat{R}Q$ (Equal angle, **A**)

$B\hat{C}A = R\hat{Q}P$ (Equal angle, **A**)

$\overline{AC} = \overline{PQ}$ (Equal side, **S**)

∴ Congruency exists between △ACB and △PQR

4}. RHS

Having Right angle Triangle,

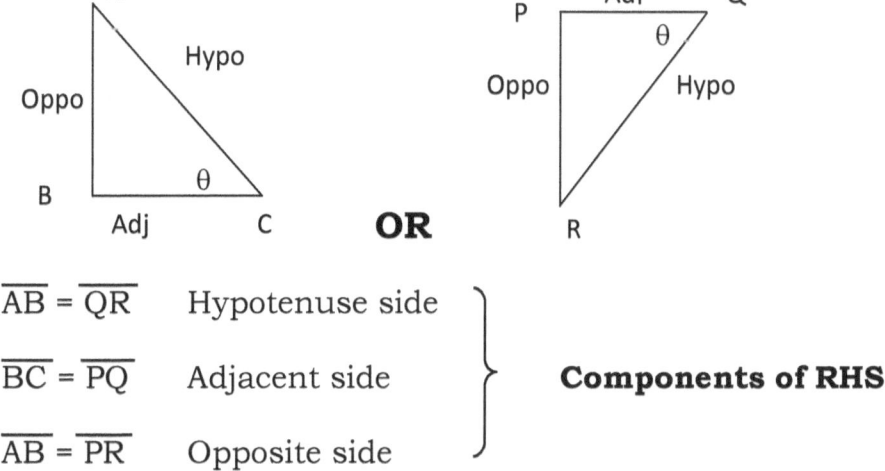

OR

$\overline{AB} = \overline{QR}$ Hypotenuse side ⎫

$\overline{BC} = \overline{PQ}$ Adjacent side   ⎬ **Components of RHS**

$\overline{AB} = \overline{PR}$ Opposite side   ⎭

∴ Congruency exists between Right Angled Triangle

$A\hat{B}C = 90^0$

SIMILARITY PROPERTIES OF TRIANGLE

Similarities between two or more triangles occur if and only if any of this condition is satisfied:

(1) EQUITRIANGULAR

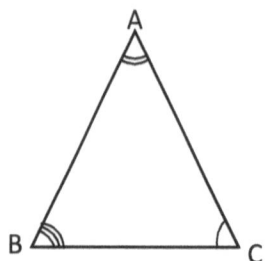 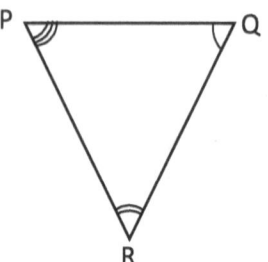

$$\left.\begin{array}{l} \overline{BC} = \overline{PQ} \\\\ \overline{BA} = \overline{PR} \\\\ \overline{AC} = \overline{RQ} \end{array}\right\} \text{Corresponding side equal} \left.\begin{array}{l} \\ \\ \end{array}\right\} \text{Equi-angular and similar for two tiangles}$$

Also, $\hat{CBA} = \hat{QPR}$ (Corr. angles equal)

$\overline{AC} = \overline{PQ}$ (*Equal angle, A*)

(2) CORRESPONDING SIDES ARE IN CONSTANT RATIO

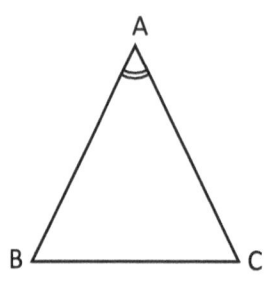 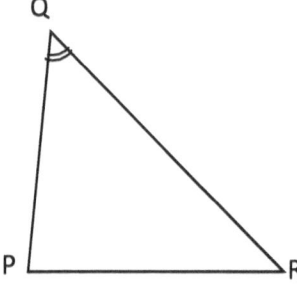

$$\frac{|AB|}{|PQ|} \quad = \quad \frac{|BC|}{|PR|} \quad = \quad \frac{|AB|}{|QR|}$$

(3) The angle of one triangle is equal to one angle of the other and their corresponding angles are in constant ratio, thus:

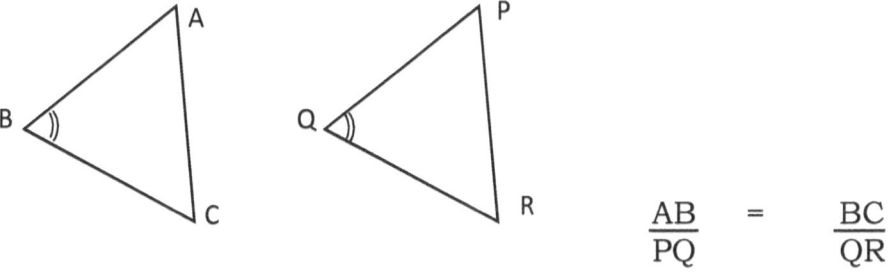

$$\frac{AB}{PQ} = \frac{BC}{QR}$$

11.02 SIMILAR TRIANGLES

THEOREM I:

A line drawn parallel to one side of a triangle divides the other side in the same ratio, thus:

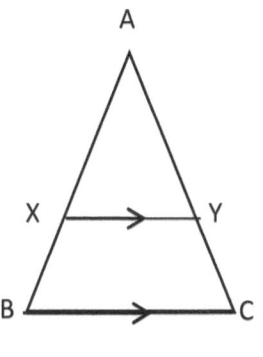

$$\overline{AX}{:}\overline{XB} = \overline{AY}{:}\overline{YC}$$

THEOREM II:

If a line divides two sides of a triangle in the same ratio, then the line is parallel to the third side of the triangle, thus:

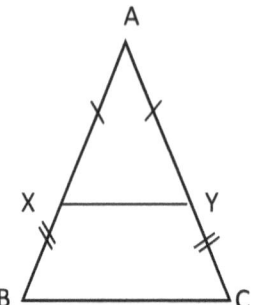

We have it that $\overline{AX}/\!/\overline{XB}$ {*meaning that line XY is parallel to line BC*}

THEOREM III:

If two triangles are equi-angular, their corresponding angles are in proportion or ratio, thus:

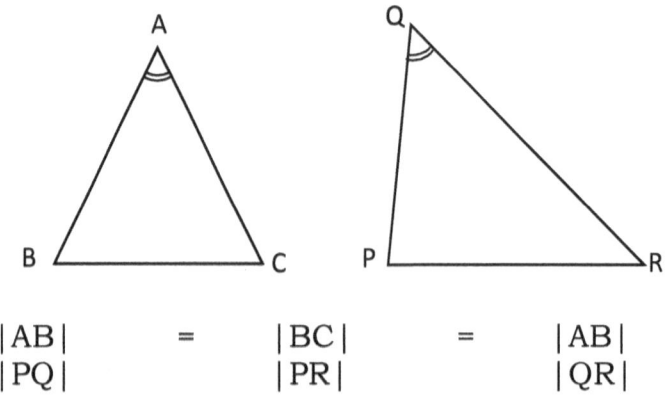

$$\frac{|AB|}{|PQ|} \quad = \quad \frac{|BC|}{|PR|} \quad = \quad \frac{|AB|}{|QR|}$$

THEOREM IV:

If the corresponding sides of two triangles are in proportion, then triangles are equi-angular and similar, thus:

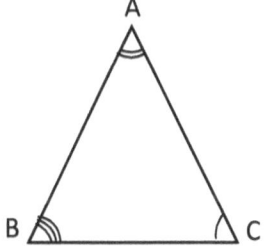 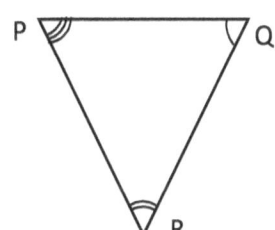

$$\left. \begin{array}{l} \overline{BC} = \overline{PQ} \\[2mm] \overline{BA} = \overline{PR} \\[2mm] \overline{AC} = \overline{RQ} \end{array} \right\} \text{Equal corresponding side}$$

Also, CBA = QPR (*Equal corresponding angles*)

11.03 QUADRILATERALS AND THEIR PROPERTIES

This is a four (4) sided polygon with four (4) angles such as squares, rectangles, rhombuses, parallelograms, trapeziums and kites.

1. SQUARES

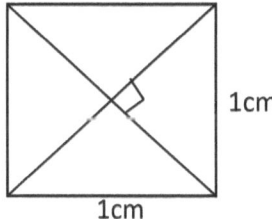

1cm

1cm

PROPERTIES

i.   All four (4) sides are equal.

ii. All four (4) angles are equal and are right angled.

iii. Equal diagonals and length.

iv. Diagonals bisect each other at right angles.

v.   Diagonal bisects the angles.

2. RECTANGLES

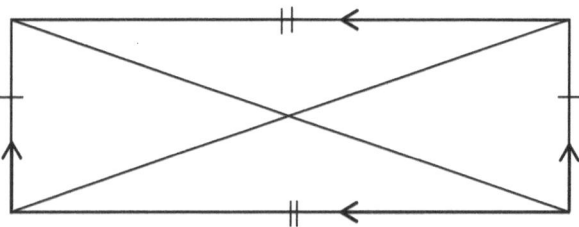

PROPERTIES

i.   All four (4) angles are equal and are right angles.

ii. Opposite sides are equal and parallel.

iii. Diagonals are equal.

iv. Diagonal do not bisects the angles.

3. RHOMBUS

PROPERTIES

i.   Slanted Square.

ii. All four (4) angles are equal.

iii. Opposite sides are equal and parallel.

iii. Opposite angles are equal.

iv. Diagonals are equal.

v. Diagonal bisects one another at right angles.

4. PARALLELOGRAM

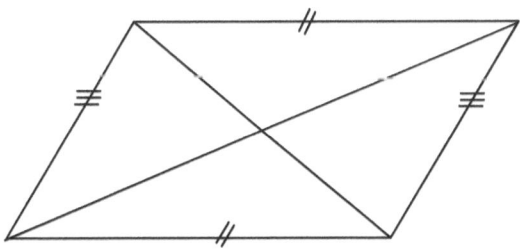

PROPERTIES

Area of parallelogram

   i. Slanted rectangle.

   ii. Opposite sides are equal and parallel.

   iii. Opposite angles are equal.

   iv. Diagonal bisects one another but not at right angles.

   v. Diagonal do not bisects the angles.

5. TRAPEZIUM

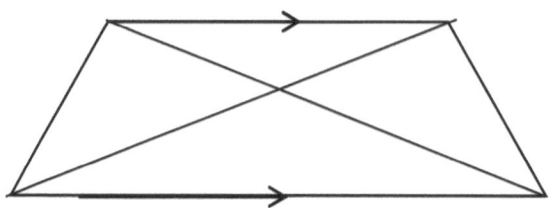

PROPERTIES

i. Has a pair of opposite side parallel.

ii. Diagonal divides the figure into two unequal triangles.

6. KITES

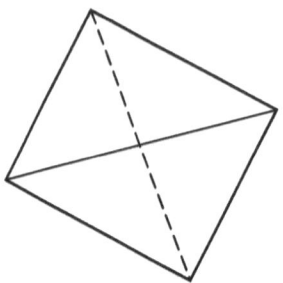

PROPERTIES

i. Diagonals are not equal (*unlike a rhombus*).

ii. Quadrilateral with a pair of adjacent sides equal.

REVISION EXERCISE 11:00

Solve for congruency ,

1.

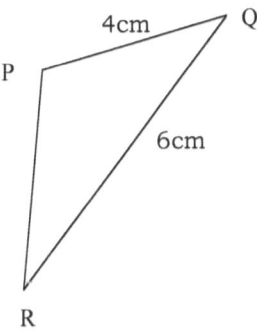
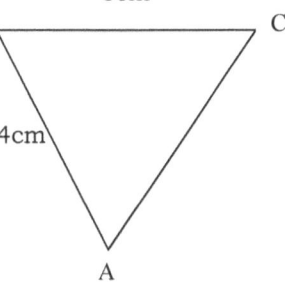

2

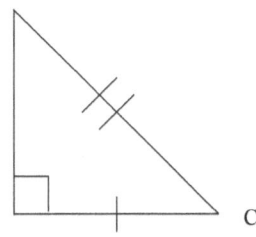
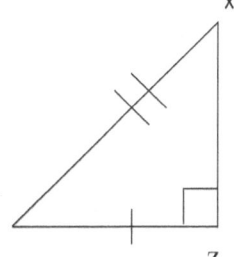

Chapter Twelve

POLYGONS

12.01 POLYGONS

THEOREM I

The sum of the interior angles in a convex polygon of n sides is:

$$(2n - 4) \, rt\angle s \qquad\qquad \text{equation (i)}$$

$$= (2n - 4) \times 90^0$$

$$(n - 2) \times 180^0$$

where n can be:

Table 1:1

No of	Name of Polygon	Interior angles	Sum of exterior angles
3	Triangle	2rt∠s	4rt∠s
4	Quadrilateral	4rt∠s	4rt∠s
5	Pentagon	6rt∠s	4rt∠s
6	Hexagon	8rt∠s	"
7	Heptagon	10rt∠s	"
8	Octagon	12rt∠s	"
9	Nonagon	14rt∠s	"
10	Decagon	16rt∠s	4rt∠s
.	.	.	.
.	.	.	.
.	.	.	.
n	n-sided polygon	(2n - 4)4∠s	4rt∠s

EXAMPLE 12.00:

The sum of an exterior angle of a regular polygon has its interior angle as same. Find the number of sides, hence, the name of the polygon.

SOLUTION:

By formula

Interior angle of regular polygon =

Sum of its exterior angle

$\therefore$ $(2n - 4)rt\angle s = rt\angle s.$

     $(n - 2)rt = 4rt$

     $(n - 2) \times 90^0 = 90^0$

     $n - 2 = 4$

     n = 6

Hence, the name of the polygon is hexagon (6 sided)

EXAMPLE 12.01:

How many sides has a regular polygon, if the sum of the interior angles is 2160^0

SOLUTION:

Sum of interior angles of a polygon is given as $(2n - 4)$rt∠s or $2(n - 2)$rt∠s

$2(n - 2)$rt∠s $= 2160^0$

$(n - 2)$rt∠s $= 1080$

$(n - 2) \times 90^0 = 1080^0$

$90^0 n - 180^0 = 1080^0$

$90^0 n = 1080^0$

n = 14

REVISION EXERCISE 12.00:

1. The size of each interior angle of a regular pentagon is 108^0

   Find the size of each exterior angle.

2. If the exterior angle of a 5-sided polygon is equal, find the exterior angle of the polygon.

3. Find the exterior angle of 8-sided polygon.

4. Find the sum of the interior angle of a polygon with 10 sides, 11 sides etc.

Chapter Thirteen

THEORY OF CIRCLES

13.01 CIRCLE:

This is a closed path traced out by point equidistance from a fixed point called the centre. It can also be defined as a locus of a point.

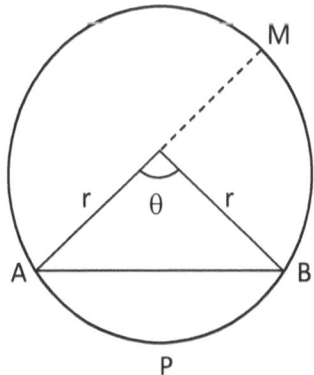

The length APB, AMB are called **ARCS** of the circle, denoted by arc APB and AMB.

APB is called **MINOR ARC** while, AMB is called **MAJOR ARC**.

Half a circle is called **SEMI CIRCLE**. Any arc subtending an angle at the centre of the circle is called the **SECTIONAL ANGLE, θ**.

A **CHORD** AB of a circle is a straight line joining two points on the circumference. The chord AOM which divide the circle into two equal halves is called **DIAMETER**, (D)

CIRCUMFERENCE is line which forms the circle or points forming the circle i.e APBM. The area covered by APB is called **MINOR SEGMENT**, while that of AMB is called **MAJOR SEGMENT.**

By formula,

Area of a circle $A = \pi r^2$

Where $\pi = {}^{22}/_7$

     $\pi = 3.142$

     $r = $ radius

     $r = {}^{D}/_2$

$\therefore$ Diameter of a circle, $D = 2r$

Circumference, $C = 2\pi r$

Length of an arc, $L_{arc} = \dfrac{\theta^0 \times 2\pi r}{360^0}$

Area of a sector, $A_{sector} = \dfrac{\theta^0 \times \pi r^2}{360^0}$

THEOREM 1:

This states that:

a) A line joining the centre of a circle to the midpoint of a chord is perpendicular to the chord, thus.

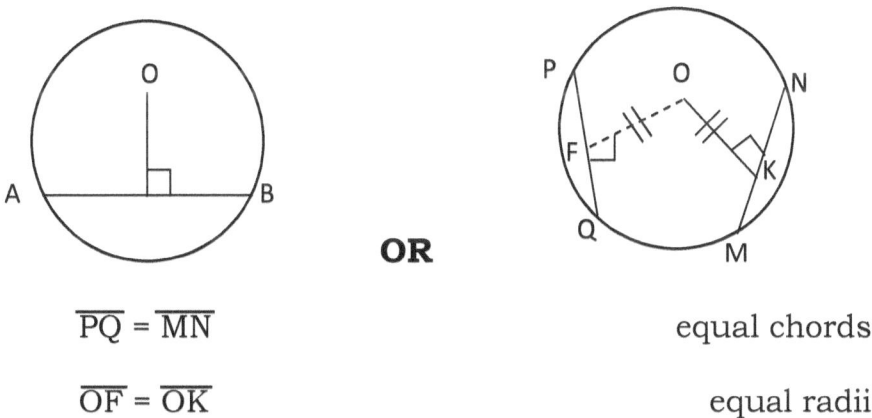

OR

$\overline{PQ} = \overline{MN}$ equal chords

$\overline{OF} = \overline{OK}$ equal radii

b) Conversely, the perpendicular drawn from the centre of a circle to a chord bisects the chord.

c) Equal chords stand on equal arcs.

THEOREM II

The angle which an arc of a circle subtends at the centre of the circle is two time that which it subtends on any part of the circumference.

Pictorially,

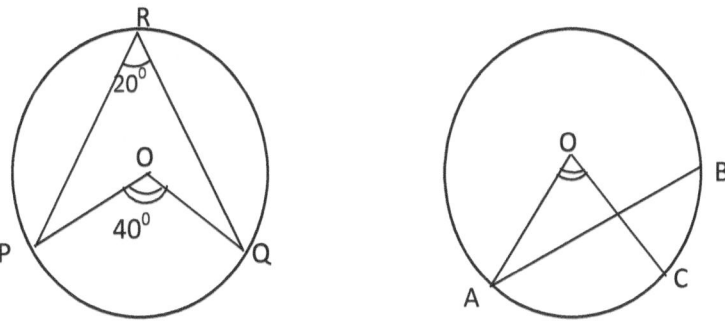

$P\hat{Q}R = 2 \times P\hat{R}Q$ and $A\hat{O}C = 2 \times A\hat{B}C$

THEOREM III

The angles subtended at the circumference by the diameter is a right angle

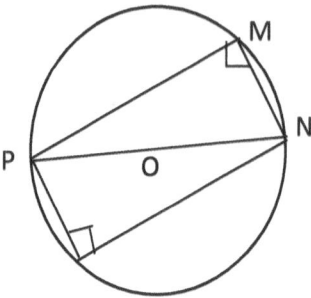

$$P\hat{M}N = P\hat{K}N = 90^0$$

THEOREM IV

If one side of a cyclic quadrilateral is produced, the exterior angle so formed is equal to the interior opposite angle.

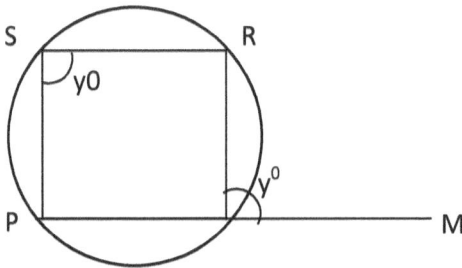

$$R\hat{Q}M = R\hat{S}P$$

The area covered by APB is called **MINOR SEGMENT**, while that of AMB is called **MAJOR SEGMENT**.

EXAMPLES ON CIRCLE THEOREMS

EXAMPLE 13.00:

BASE ANGLE OF AN ISOSCELES TRIANGLE

1. Given a circle, thus:

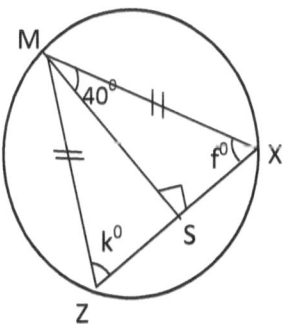

$\triangle X\hat{S}M = 180^0$ (Right angled triangle)

$S\hat{M}X + X\hat{S}M + M\hat{X}S = 180^0$ (*Sum of angles in a triangle*)

MXS = SZM (*Base angle of an isosceles triangle*)

$= k^0 = f^0$

So, $f^0 = 90^0 - 40^0 = \textbf{50}^0 = \textbf{k}$

2. Given a circle as seen below, find PRS

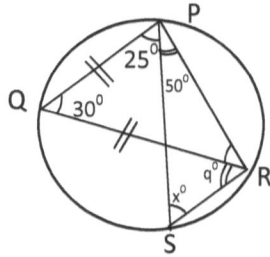

Since, RQP = PSR

= 30^0 = x^0 (Angle in the same segment)

But $30^0 + 50^0 = x + 50^0 = 80^0$

∴ $80^0 + 75^0 + q^0 = 180^0$ (*Summation of* △ *SPR*)

⇒ $155^0 + q^0 = 180^0$

$q0 = 1800 - 1550 = 25^0$

⇒   PSR = $75^0 + 25^0$ = **100^0**

REVISION EXERCISE13.00:

{1} An arc ACB subtend an angle of 64^0 at the centre of diameter 25cm.Find the { a} the length of the chord AD { b} the perimeter of the segment ACD { Use pie = 3.142}

{2} A chord of a circle of a radius 9.5cm is 5cm from centre ,What is the length of the chord ?

Chapter Fourteen

BOOLEAN ALGEBRA AND LOGICAL REASONING.

1.1 BOOLEAN ALGEBRA AND LOGICAL REASONING

In the mid 1800's George Boole developed a new type of arithmetic logic that today bears his name: **BOOLEAN ALGEBRA:** A system of Mathematical logic which uses symbols and set theory to represent logical operations in Mathematical form. This allows us to solve complex Mathematical and logical problems by the manipulation of only two (2) conditions, thus: true (1) false (0) its logic basically requires only three basic logic blocks or gates: **AND, OR** and **NOT**. Each gate normally performs some simple functions (operations) which make us call them **LOGIC ELEMENTS.**

See more in *Simplified Mathematics Series II for Secondary Schools with Answers.*

Chapter Fifteen

AREA AND SOLUTIONS OF TRIANGLES

15.01 HERON'S (*HERO'S*) FORMULA:

A formula connecting the area of a triangle with its sides. It is a method of solution of triangle. Given a triangle A, B, C with sides a, b and c.

The triangle consists of (i) three sides, (ii) three angles (ii) three apexes etc.

Given any of the angles and any side, others can be found by the use of Sine and Cosine rule while Pythagoras theorem can be used if it is a right angled triangle.

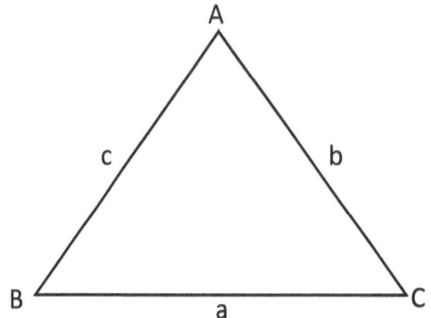

The area, A of the triangle can be found, thus

$$s = \tfrac{1}{2}(a + b + c) \quad \Rightarrow \quad (\textit{Semi Perimeter})$$

Area of the $\triangle$, $A = \sqrt{s(s-a)(s-b)(s-c)}$

EXAMPLE:

Find the area of a triangle, given

a = 2cm b = 3cm and c = 4cm

by formula of Hero's

$$s = \frac{a + b + c}{2}$$

$$= \frac{2 + 3 + 4}{2} = {}^9/_2 = 4.5$$

Area, $A = \sqrt{4.5\ (4.5\text{-}2)\ (4.5\text{-}3)\ (4.5\text{-}4)}$

$$= \sqrt{4.5\ (2.5)(1.5)(0.5)}$$

$$= \sqrt{8.4375}$$

$$= \mathbf{2.904cm^2}$$

15.02 SINE AND COSINE RULES

SINE RULES

Given a triangle A,B,C thus;

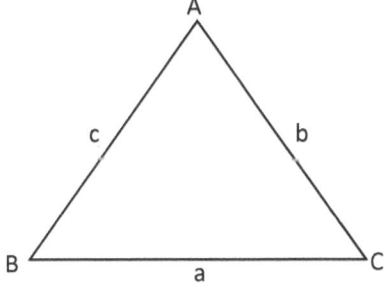

Sine Rule states that;

$$^a/_{\text{Sin A}}^{\text{o}} = ^b/_{\text{Sin B}}^{\text{o}} = ^c/_{\text{Sin C}}^{\text{o}}$$

Any two of these can be combined to find a desired variable

EXAMPLE

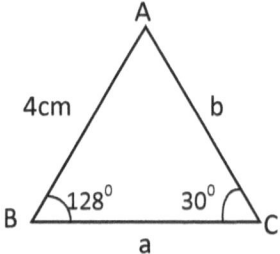

Find (i) A^o (ii) side b

SOLUTION

Combining $^c/_{\text{Sin C}}^{\text{o}} = ^a/_{\text{Sin A}}^{\text{o}}$

Substituting $^4/_{\text{Sin 30}}^{\text{o}} = ^3/_{\text{Sin A}}^{\text{o}}$

Cross multiplying,

$4 \text{ Sin } A^o = 3 \text{ x Sin30}^o$

$4 \text{ Sin } A^o = 3 \text{ x } 0.5$

$\text{Sin } A^o = 3 \text{ x } 0.5/4$

$= 0.375$

$A^o = \text{Sin}^{-1} 0.375$

$= 22.02^0$

$= 22^0$ (*to the nearest degree*)

(ii) Combining $b/_{Sin\ B}{}^0 = a/_{Sin\ A}{}^0$

Substituting $b/_{Sin\ 128}{}^0 = 3/_{Sin\ 22}{}^0$

Cross multiplying,

b Sin 22^0 = 3 x Sin 128^0

b Sin 22^0 = 3 x 0.7880

$b = 3\ x\ \dfrac{0.7880}{Sin\ 22^0}$

$= 3\ x\ \dfrac{0.7880}{0.3746}$

= 6.31 i.e   **b = 6m**

COSINE RULE

In a right angled triangle,

$a^2 = b^2 + c^2$, but for a non right angled triangle, it can be extended thus;

$a^2 = b^2 + c^2 - 2bc\ Cos\ A^0$ equation (i)

$b^2 = a^2 + c^2 - 2ac\ Cos\ B^0$ equation (ii)

$c^2 = a^2 + b^2 - 2ab\ Cos\ C^0$ equation (iii)

All the above can be used to solve problems when:

(i) Given two sides and the included angle

(ii) Given two sides and a non included angle

(iii)   Given two angles and one side of the triangle.

EXAMPLE

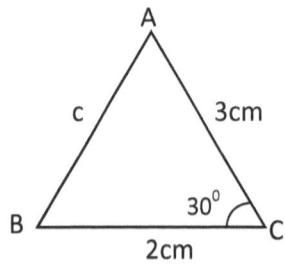

$a = 2$, $b = 3$ and $C^0 = 30^0$

Using the equ (iii)

c^2 $= a^2 + b^2 - 2ab \cos C^0$

$= 2^2 + 3^2 - 2 \times 2 \times 3 \cos 30^0$

$= 4 + 9 - 12 \times 0.8660$

$= 13 - 10.392$

c^2 $= 2.60$

c $= \sqrt{2.60}$ $= \mathbf{1.61cm}$

15.03 <u>AREA OF TRIANGLES</u> *{continued}*

Given an isosceles triangle, thus

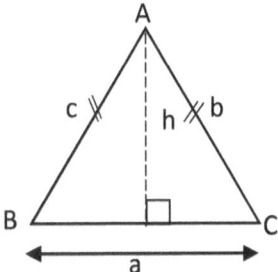

Area, A = ½ b x h ,where b = base , h = height

We see that Sin B^0 = Opp. (**SOHCAHTOA**)
            Hypo.

∴   Sin B^0 = h
                    c , h = **c Sin B^0**

Area of △   = ½ a x h

            = ½ a x c Sin B^0

            = ½ ac Sin B^0 equation. (1)

Similarly others are:

            = ½ ab Sin C^0 equation.(2)

            = ½ ac Sin B^0 equation.(3)

NOTE:

SOHCAHTOA

Sinθ = Opp./Hypo. , **C**osθ = Adj./Hypo, **T**anθ = Opp./Adj.

Chapter Sixteen

ANGLE OF ELEVATION AND DEPRESSION

16.01 ANGLE OF ELEVATION AND DEPRESSION:

ANGLE OF ELEVATION

This is the angle between the horizontal plane and the oblique line from the observer's eye to some object above his eye.

Diagrammatically, the figure

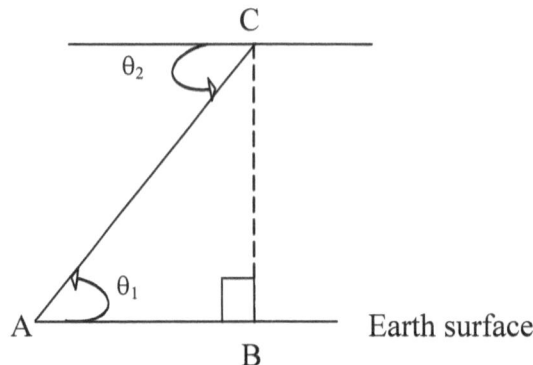

Having the angle of elevation as θ_1

$\overline{BC}$ = height, $\overline{AB}$ = Earth surface

$$\text{Tan } \theta_1 = \frac{BC}{AB}$$

ANGLE OF DEPRESSION

And θ_2 = *angle of depression*, as it involves looking down from a particular height. It also means the angle between the horizontal plane and the oblique line joining the observer's eye to some object beneath the line of is eye.

EXAMPLE 16.00: A vertical mast is erected 24m away from a building. From the top of the building the angle of elevation and depression of the top and foot of the mast are 60^0 and 45^0 respectively.

Calculate:

i. The height of the mast

ii. The distance between the top of the building and the top of the mast

(*Correct to nearest meters* **NECO 2009**.}

Chapter Seventeen

BEARING AND LOCATIONS

17.01 BEARING AND LOCATIONS

This is the clockwise angular relationship between two distances. It is also defined as the bearing of a line (surveying). The angle which the line makes with north and south line; its direction relative to the north south line is also called bearing.

EXAMPLE 17.01:

The bearing 045^0 or $N45^0$ E which means face north and turn 45^0 towards East, can be drawn as

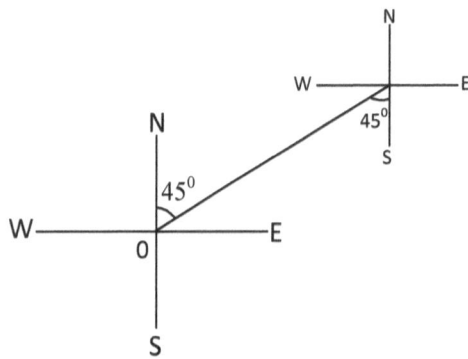

EXAMPLE 17.02:

The bearing 030^0 or $S30^0$ W means face south and turn 30^0 towards west, can be drawn as

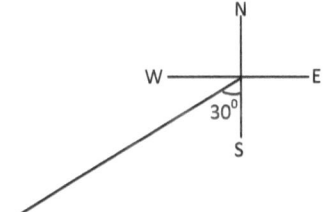

Also, we have the geographical notation or 3 - digit notations such as 000⁰, 090⁰, 270⁰, etc

EXAMPLE 17.03:

A traveler, moves from a town P on a bearing of 055⁰ to a town Q, 200km away. He then moves from Q on a bearing of 155⁰ to a town R, 400 km from Q, find the

i.   Distance between P and R

ii. Bearing of P from R (*correct to the nearest degree*)

SOLUTION

Diagrammatically

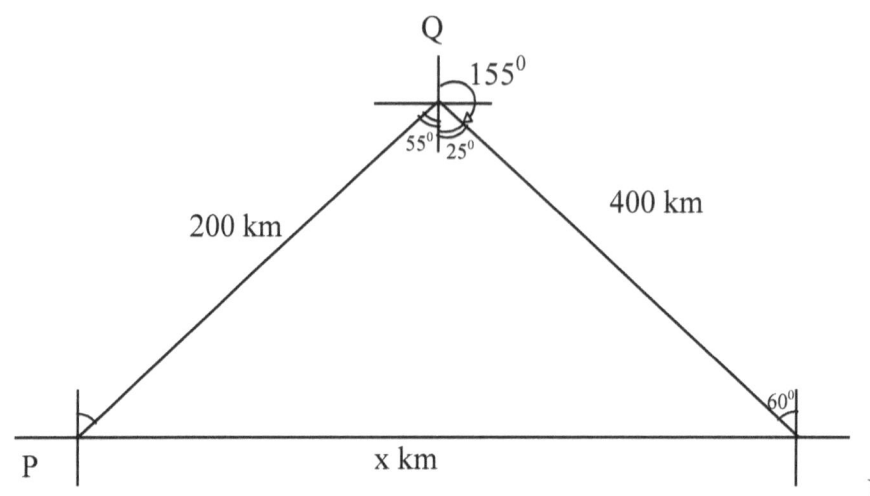

Illustrating further,

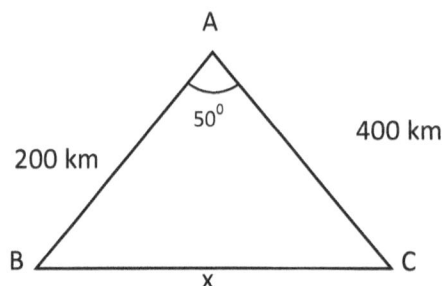

Using cosine rule,

$$a^2 = b^2 + c^2 - 2bc \, \text{Cos} \, A^0$$

$$= (400)^2 + (200)^2 - 2(400)(200) \, \text{Cos} \, 80^0$$

$$= 160,000 + 40,000 - 160,000 \times 0.1736$$

$$= 20,000 - 27783.708$$

a^2 $= 172216.29$ (*Square rooting,*)

a $= 414.9$ km

∴   a   = **415 km** distance xkm between P and R

(ii) Using the suitable Sine rule

$$\frac{b}{\text{Sin B}^0} = \frac{a}{\text{Sin A}^0}$$

$$\frac{400}{\text{Sin B}^0} = \frac{415}{\text{Sin 80}^0}$$

Cross multiplying

Sin B^0 $= \dfrac{400 \times 0.9848}{415}$

Sin B^0 0.9492

B^0 $= \text{Sin}^{-1} 0.9492$

$= 71.66^0$

$= \mathbf{72^0}$

$72^0 + 055^0 + 18^0 = \mathbf{307^0}$

The bearing of P from R correct to the nearest degree is

307⁰

EXAMPLE 17.04:

Three towns P,Q and R are such that the distance between P and Q is 50km and the distance between P and R is

90km. if bearing of Q from P is 075⁰ and the bearing of R from P is 310⁰, find the

(i) Distance between Q and R

(ii) Bearing of R from Q

SOLUTION

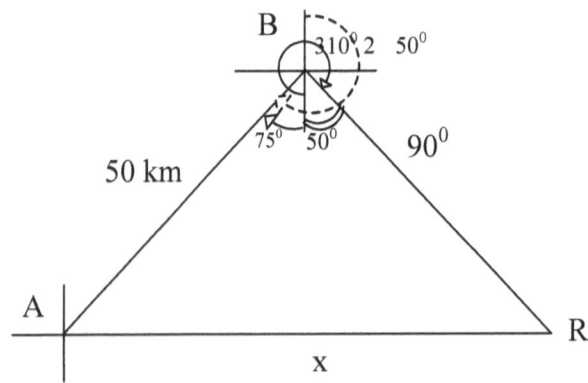

(i) Using Cosine Rule, we find

$$b \quad = x \text{ as}$$

$$b^2 \quad = a^2 + c^2 - 2ac \text{ Cos B}^0$$

$$= 90^0 + 50^0 - 2 \times 90 \times 50 \text{ (Cos } 125^0)$$

$$= 8100 + 2500 - 9000 \text{ (-0.5735)}$$

$$= 10600 - 5161.5$$

$$= 15761.5$$

$$b^2 \quad = 15761.5$$

$\therefore$ b $= \sqrt{15761.5}$ = **125.5**

        = **126km**

(ii) To find the bearing of R from Q

It implies the addition of 180^0 to 75^0

        i.e $180^0 + 75^0$ = **255^0**

{as shown by the dotted lines in the diagram above.}

EXAMPLE 17.05:

The bearing of Q from P is 122^0 what is the bearing of P from Q?

SOLUTION

Diagrammatically

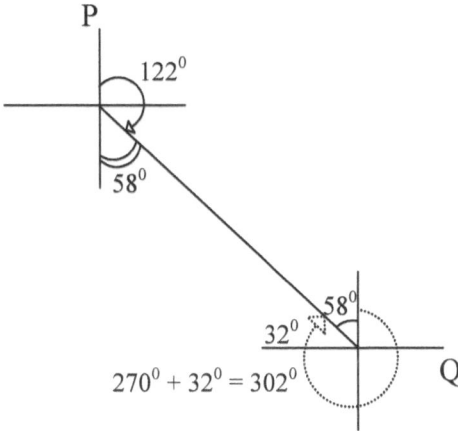

∴ $270^0 + 32^0 =$ **302^0** this is the bearing of P from Q.

EXAMPLE 17.06:

Two men P and Q set off from a base camp R prospecting for oil. P moves 20km on a bearing of 205^0 and Q moves 15km on a bearing of 60^0

Calculate   (i)   The distance of Q from P

             (ii)   The bearing of Q from P

(*correct to nearest whole number*)

SOLUTION

Diagrammatically

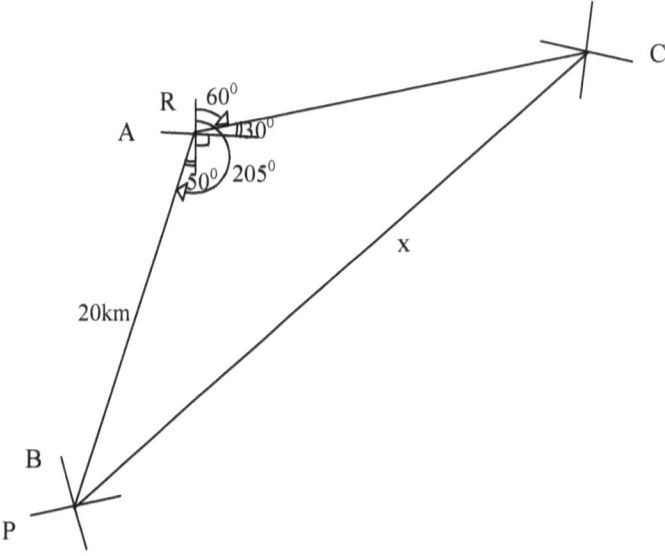

∴ $25^0 + 90^0 + 30^0 = 145^0$

⇒ $A^0 = 145^0$

(i) Using the Cosine Rule, distance, x

⇒ $a^2 = b^2 + c^2 - 2bc \, Cos \, A^0$

      $= 15^2 + 20^2 - 2 \times 15 \times 20 \, (Cos \, 145^0)$

      $= 225 + 400 - 600 \, (-0.8191)$

      $= 625 - -491.46$

      $a^2 = 1116.46$

∴ $a = \sqrt{1116.46}$

      = 33.41

      = 33km

(ii) Using the Sine Rule

$$\frac{a}{Sin \, A^0} = \frac{b}{Sin \, B^0}$$

$$\frac{33.41}{Sin \, A^0} = \frac{15}{Sin \, B^0}$$

Cross multiply

    $33.41 \, Sin \, B^0 = 15 \, Sin \, 145^0$

        $Sin \, B^0 = \dfrac{5 \times Sin \, 145^0}{33.41}$

        $= \underline{15 \times 0.57.35}$

$$33.41$$

$$= \mathbf{0.25748}$$

$B^0 \quad = Sin^{-1} 0.25748$

$$= 14.920$$

$$= \mathbf{15^0}$$

∴   The bearing of Q from P is

$25^0 + 15^0 = \mathbf{40^0}$

REVISION EXERCISE 17.00

1. An airplane leaves airport A and flies on a bearing 035^0 for 1¼ hours at 600km per hour to airport B. it then flies on a bearing 130^0 for 1½ at 400 km per hour to airport, C calculate the

    i. distance from C to A

    ii. The bearing of C and A

(correct to the nearest whole number)

Chapter Eighteen

SEQUENCE AND SERIES

18.01 SEQUENCE AND FIBONACCI SEQUENCE

SEQUENCE

This is defined as a set of quantities each of which is formed from one or more of the proceedings according to some fixed laws. Thus:

(a) 1,3,5,7,9,11,13 . . .

(b) 0,3,6,9,12,15,18 . . .

(c) 0,-4,-8,-12 are examples.

(d) $1, 1/2, 1/3, 1/4, 1/5, 1/6$ . . .

In (a) each term is formed by adding 2 to the preceding terms;

In (b) is by adding 3 to the preceding terms

In (c) is by adding-4 to the preceding terms,

In (d) the nth term is $1/n$.

However, the common difference in (a) = 2, (b), (c) =-4 and (d) =$-1/2$

EXAMPLE 18.00: In the sequence :2, 4, 6 ,8, 10 . . . there is an obvious pattern. Such pattern can be expressed in terms of the nth term of the sequence. In this case, the nth term is 2n.

To find the first term , put n=1 into the formula and to find the fourth terms replace the n's by 4's .say to say 4th term is 2x4 = 8.

EXAMPLE 18.01: What is the nth term of the sequence 2, 5 ,10, 17, 26 . . . ?

To find the answer, create a table thus:

n =	1	2	3	4	5
n^2 =	1	4	9	16	25
$n^2 + 1$ =	2	5	10	17	26

This the required sequence so that nth term is n^2+1.

EXAMPLE 18.02: Given 2, 6, 12, 20, 30, 42 . . . find the nth term of the sequence.

n = 1 ,2 ,3 ,4 ,5

n{ n+1 }/2 = 1, 3 ,6 , 10 , 15

2n{ n+1 }/2 = n{ n+1} = 2

→ For n =1; 1{1+1 } = 1{ 2} =1x2 = 2

n = 2; 2{ 2+1} = 2{ 3} = 2x3= 6

n= 3; 3{ 3+1} = 3{ 4} = 3x4= 12

n=4; 4{ 4+1}= 4{ 5}= 4x5=20

n=5; 5{ 5+1} = 5{ 6} = 5x6 = 30

n=6 ; 6{ 6+1 }= 6{ 7} = 6x7 = 42

So the nth term of the sequence is n { n+1}.

FIBONACCI SEQUENCE

Given: 1, 1, 2, 3, 5, 8, 13, 21 . . .

The next term of this well known sequence is found by adding together the two previous terms; thus

21 = **8+13**

13 = **8 +5**

8 = **5 + 3**

5 = **3 + 2**

3 = **2 + 1**

2 = **1 + 1**

18.02 SERIES

Given a sequence; 0+3+6+9+12+ . . . up to n^{th} term, we represent the first term with $N_1{}^{th}$, the second with $N_2{}^{th}$ by $N_3{}^{th}$ etc.

Therefore, the terms of a sequence are added, the resulting expression is called a SERIES.

It is the sum of the terms of a sequence. Finite sequence and series have defined first term and last term whereas infinite sequence and series continue indefinitely.

In Mathematics, given an infinite sequence of number $\{a_n\}$,a series is informally the result of adding all those terms together such that a_1, a_2 ,a_3 . . . . a_n can be written as more compactly using the summation symbol $\{\sum\}$.

An example is the famous series from zero's dichotomy:

$$\sum 1/2^n = 1/2 + 1/4 + 1/8 + \ldots + 1/2^n$$

The above is an infinite series.

Unlike finite summation, infinite series need tools from mathematical analysis to be fully understood and manipulated.

RELATIONSHIP BETWEEN SEQUENCE AND SERIES

DIFFERENCE BETWEEN SEQUENCE AND SERIES:

If 45, 25 , 5 . . . is an example of sequence. The sum of the numbers is series. Thus, 45+25+5+ . . . is a series.

Given $1, 1/2, 1/4, 1/8, 1/16$ . . . as a sequence, then 1+ ½ + ¼ +1/8+ 1/16 + . . . is a series. Also,1,1/2,1/3,1/4,1/5, 1/6 . . . is a sequence and the nth term is $1/n$. The two are often confused because they are closely related to each other.

NUMBER SEQUENCE: This does not imply sum of the values of the terms rather it is a list of numbers.

SERIES: This implies the sum. It is the sum of numbers.

However a series in turn can be associated with the sequence of its terms for instance, $1+1/4 + 1/16 + 1/64 + \ldots$ can be associated with $1, 1/4, 1/16, 1/64 \ldots$ That is to say without the addition{ +} signs. But may also

be associated with another sequence called the "Sequence of partial sums" thus $1, 5/4, 21/16, 85/16 \ldots$ in which the nth term of the sequence is the sum of the first n terms of the series, so we have

$$1 = 1$$

$$5/4 = 1 + \tfrac{1}{4}$$

$$21/16 = 1 + 1/4 + 1/16$$

$$85/16 = 1 + \tfrac{1}{4} + 1/16 + 1/64 \text{ etc.}$$

Then it is not too hard to see that the limit of this sequence is equivalent to the sum of the series.

ARITHMETIC PROGRESSION (A.P)

This is defined as a series from the preceding by addition to a constant quantity (common difference, d) which is formed by substituting any term from the term preceding it thus:

(i) 8, 19, 30, 41   d = 11

(ii) 4, 9, 14, 19, 24   d = 5

Arithmetic Progression (A.P) in its standard form as follows: a, a+d, a+2d . . . in which the first term is "a" and the common difference is "d".

Notice that the coefficient of d in any term is one less than the series thus the second term is a +d; the third term is a+2d; fourth term is a+3d etc. Generally the nth term is { a + (n-1)d }

Where n = number of terms

N^{TH} TERM OF AN A.P:

EXAMPLE 18.03:

Find the 7th and 24th terms of the series 21,18,15 . . .

SOLUTION

To check for d, the common difference such that

$$18 - 21 = 15 - 18 \Rightarrow -3 = -3$$

∴   the common difference, d =-3

Applying the formula, N^{th} = a +(n - 1)d

∴ 7th Term = 21 + (7 - 1) x-3

= 21 + 6 x-3

= 21 +-18 = **3**

For the 24th term

= 21 + (24 - 1) x (-3)

= 21 + 23 x (-3)

= 21 +-69

=-**48**

EXAMPLE 18.04:

Finding the Nth term of the sequence

3, 9,15, 21

SOLUTION

To check for, d

9 - 3 = 15 - 9

6 = 6

Common difference, d = 6

NOTE: a = 3

d = 6

n = unknown

Applying the formula

N^{th} = a + (n-1) d

= a + nd - d

= 3 + n (6)-6

$$= 3 + 6n - 6$$

Factorizing,

$$= 6n - 6 + 3$$

$$= 6n - 3$$

N^{th} $= 6n-3$

REVISION EXERCISE 18.00:

1. Finding the 8th and 20th terms of the series

   (a) 4, 9, 14 . . .

   (b) a, 4a, 7a . . .

2. Finding the common difference, d and the 7th term of series

   (a) x-2y, x-y, x . . .

   (b) c, c-2d, c-4d . . .

SUM OF N TERM OF A.P

Given the sum of n term of the series as

a, a+d, a+2d a+2d, a+3d . . . we denote the sum of n terms, let l denote the last term then the last term but one is l-d, the last term but two is l-2d, the last term but three is l-3d etc.

Hence, S = a + (a+d) + 9 (a+2d)+   (1-3d) +(1-2d)+(1-d) +1

Writing the series in reversed order, starting with 1

S = 1 + (1 - d) + (1-2d) + (1-3d) +   (a+3d) + (a+2d) + (a+d) + a

From these result, adding corresponding terms we have

2S = (a+l) + (a+l) + (a+l) + the bracket being repeated n times

2S = (a +l) + (a+l) + (a+l) +   n

$S = \dfrac{n\{a +l\}}{2}$

But l = a+(n-1) d

$$S = n/2 \; 2a+(n-1)d$$

EXAMPLE 18.05:

Find the last term and the sum of 25 terms of the series

12, 9, 6 . . . .

SOLUTION

Having a = 12, 1 = 6, n = 3

We then say l = a + (n-1)d

$$6 = 12 + (3-1)d$$

$$6 = 12 + 2d$$

→ $6 - 12 = 2d$

$2d = -6$

$d = -\dfrac{6}{2} = \mathbf{-3}$

OR

Is $9 - 12 = 6 - 9$

Yes it is $-3 = -3$

$d = -3$

Applying the equation (ii)

$l = 12 + (25-1)-3$

$= 12 + 24 \text{ x} - 3$

$= \mathbf{-60}$

Also using equation (iii)

$S \quad = \dfrac{25}{2} \{2(12) + (25-1)(-3)\}$

$= \dfrac{25}{2} \{24 + 24(-3)\}$

$= \dfrac{25}{2} (24 - 72)$

$= \dfrac{25}{2} (-48)$

$= 12 \frac{1}{2} \text{ x } (-48)$

$= \mathbf{-600}$

REVISION EXERCISE 18.01:

1. Given $l = -88$, $n = 16$, $s = -448$, find "a" and "d"

2. How many terms of the series 42, 39, 36 . . . must be taken that the sum may be 312?

Find: (a) The first term

   (b)   The common difference

   (c)   The sum of the first 20 terms

3. Given a series 1, $^4/_3$, $^5/_3$ as an A.P. find:

   (a) 10th term

   (b)   Sum up to the 16th term { **NECO 2009**}

GEOMETRIC PROGRESSION (G.P)

This is defined as a series of which each term in formed from the preceding by multiplying it by a constant. That is to say, dividing any term by the term preceding it, this

   (i) 1, 4, 16, 64 . . . r = 4

   (ii) 1,3,9,27 . . . r = 3 are example

THE N^{TH} TERM OF G.P

The standard form of a G.P is a, ar, ar^2, ar^3, ar^4 . . . ar^n

Thus, 3rd term is $ar^{(3-1)} = ar^2$

   4th term is $ar^{(4-1)} = ar^3$ etc.

n^{th} term is **ar^{n-1}**

EXAMPLE 18.06:

Finding the 10th term of the G.P, 16,-8, 4,-2 . . . .

SOLUTION

$a = 16$

To check for, r Is$-^8/^{16} = ^4/_{-8}$?

$\Rightarrow$ $-^1/_2 = -^1/_2$

$r = -^1/_2$

So by the formula, $N^{th} = ar^{n-1}$

$\Rightarrow$ $10^{th} = 16(-^1/_2)^{10-1}$

$= 16(-^1/_2)^9$

$= 0.0312$

EXAMPLE 18.07:

Finding the 8th term of the series$-^1/_3$, $^1/_2,-^{3/4}$ . . . .

SOLUTION for r

Is $\quad$ $\frac{1}{2} \div \frac{1}{3} = -\frac{3}{4} \div -\frac{1}{2}$?

$\Rightarrow \quad \frac{1}{2} x \cdot \frac{-3}{1} = -\frac{3}{4} x \cdot \frac{-2}{1}$

$\Rightarrow \quad \frac{3}{-2} = -\frac{3}{2}$

$\therefore \quad r = -\frac{3}{2}$ and $a = -\frac{1}{3}$

By the formula $N^{th} = ar^{n-1}$

$\quad 8^{th} = \left(-\frac{1}{3}\right) x \left(-\frac{3}{2}\right)^{8-1}$

$\quad = \left(-\frac{1}{3}\right) x \left(^{-3}/_{2}\right)^{7}$

$\quad = \frac{1}{3} x (1.5)^{7}$

$\quad = \frac{1}{3} x\ 17.0859 = 5.0953 = \mathbf{5.7}$

18.03 <u>GEOMETRIC MEAN</u>

When three quantities are in Geometric Progressions, the middle is called the Geometric Mean between the other two. Finding this in a given quantities say "a" and "b". Let G be the required mean, then since a, G, b are in Geometric progression.

$\quad \dfrac{b}{G} = \dfrac{G}{a} = r$, the common ratio

$\therefore$ G^2 = ab or G = $\sqrt{ab}$

EXAMPLE18.08: Given 2, G, 8 . . . as a G.P find the Geometric mean

SOLUTION

In 2,G, 8 the G $= \sqrt{ab}$

$$= \sqrt{2 \times 8}$$

$$= \sqrt{16}$$

$$= 4$$

because $^4/_2 = {}^8/_4 \Rightarrow 2 = \mathbf{2}$

So, we have 2, 4, 8 as the series.

EXAMPLE18.09: Find the Geometric mean of ½ and $^1/_8$

SOLUTION

$$G = \sqrt{ab}$$

$$= \sqrt{\tfrac{1}{2}} \times \sqrt{^1/_8} = \sqrt{^1/_{16}} = \tfrac{1}{4} \text{ because } \tfrac{1}{2}, \tfrac{1}{4}, {}^1/_8$$

Verifying that $^1/_8 \div \tfrac{1}{4} = \tfrac{1}{4} \div \tfrac{1}{2}$

$$\Rightarrow \quad \tfrac{1}{2} = \tfrac{1}{2}$$

SUM OF N TERM OF G.P

This is given by $S_n = a \dfrac{(1-r^n)}{1-r}$ if $r < 1$ **OR**

$$S_n = a \dfrac{(r^n-1)}{r-1} \text{ if } r > 1$$

where a = first term

$\quad$ r = common ratio

$\quad$ n = number of terms in G.P

EXAMPLE 18.10: find the sum of the fourth 4th and 8th terms of a G.P given as

$$^1/_3, \, ^1/_9, \, ^1/_{27} \cdots$$

SOLUTION

In the above G.P, a = $^1/_3$ and r = is $^1/_9 \div ^1/_3 = ^1/_{27} \div ^1/_9$

$\Rightarrow \quad ^1/_9 \times ^3/_1 = ^1/_{27} \times ^9/_1$

$\quad ^1/_3 = ^1/_3$

$\mathbf{r = {}^1/_3 = 0.33}$

Thus the sum of the first 4 terms is S_4

$\Rightarrow \quad S_4 = {}^1/_3 \dfrac{(1-(^1/_3)^4)}{1-^1/_3}$ since $r < 1$

$$S_4 = \frac{1}{3}\frac{(1-\frac{1}{81})}{\frac{2}{3}}$$

$$= \frac{1}{3}\left(\frac{80}{81} \div \frac{2}{3}\right)$$

$$= \frac{1}{3}\left(\frac{80}{81} \times \frac{3}{2}\right)$$

$$= 0.33 \times 0.4949$$

$$= \mathbf{0.4888}$$

For sum of the first 8th term,

$$S_8 = \frac{a(1-r^8)}{1-r}$$

$$= \frac{\frac{1}{3}(1-(\frac{1}{3})^8)}{1-\frac{1}{3}}$$

$$= \frac{\frac{1}{3}(1-1/6561)}{1-\frac{1}{3}}$$

$$= \frac{1}{3}\left(\frac{6560}{6561} \div \frac{2}{3}\right)$$

$$= \frac{1}{3}\left(\frac{6560}{6561} \times \frac{3}{2}\right)$$

$$= \frac{1}{3}\left(\frac{19680}{13122}\right)$$

$$= \frac{1}{3}(1.4997) = \mathbf{0.4992}$$

18.04 <u>SUMS TO INFINITY</u>

To find the sum to infinity of a G.P such as $^1/_3, ^1/_9, ^1/_{27} \cdots$

SOLUTION

$$S = \frac{a}{1-r} = \frac{^1/_3}{1-^1/_3}$$

$$= \frac{^1/_3}{^2/_3}$$

$1/3 \div 2/3 = {}^{1/3} \times {}^{3/2} = \mathbf{1/2}$

∴ Sum of the G.P is = 0.5 as n ➔ ∞

SUM OF A GEOMETRIC PROGRESSION TO INFINITY

To find the sum to infinity of a G.P

EXAMPLE 18.11:

Given the series $^1/_3, ^1/_9, ^1/_{27} \cdots$

Find the

(a) 4th term

(b) 8th term

(c) Sum of the first 4 terms

(d) Sum of the first 8 terms

(e) Sum of the whole G.P

SOLUTION:

In the given G.P, $a = \frac{1}{3}$ and $r \rightarrow \frac{1}{9} \div \frac{1}{3} = \frac{1}{27} \div \frac{1}{9}$

$$\frac{1}{9} \times \frac{3}{1} = \frac{1}{27} \times \frac{9}{1}$$

$$\frac{1}{3} = \frac{1}{3}$$

$$r = \frac{1}{3}$$

By formula,

$\rightarrow$ Nth term = ar^{n-1}

(a) 4th term = ar^{4-1}

$$= \left(\frac{1}{3}\right)\left(\frac{1}{3}\right)^3$$

$$= \frac{1}{3}\left(\frac{1}{27}\right)$$

$$= \frac{1}{81}$$

(b) 8th term = ar^{8-1}

$$= \left(\frac{1}{3}\right)\left(\frac{1}{3}\right)^7 = \frac{1}{6561}$$

(c) The sum of the first 4 term is (S_4)

➔ $S_4 = \dfrac{a(1-r^n)}{1-r}$ if $r < 1$ equation. (i)

OR

$S_4 = a\dfrac{(r^n-1)}{r-1}$ if $r > 1$ equation. (ii)

Using equation. (i), where $r < 1$, because

$r = {}^1/_3$

$S_4 = \dfrac{{}^1/_3 (1 - ({}^1/_3)^4)}{1-{}^1/_3}$

$= \dfrac{{}^1/_3 (1 - {}^1/_{81})}{{}^2/_3}$

$= {}^1/_3({}^{80}/_{81} \div {}^2/_3)$

$= {}^1/_3({}^{80}/_{81} \times {}^2/_3) = \textbf{0.4938}$

(d) Sum of the first 8 terms is S_8

$S_8 = \dfrac{a(1 - r^8)}{1 - r}$

$= \dfrac{{}^1/_3\left[1 - ({}^1/_3)^8\right]}{1-{}^1/_3}$

$= {}^1/_3({}^{6560}/_{6561} \div {}^2/_3)$

$= {}^1/_3({}^{6560}/_{6561} \div {}^3/_2) = \textbf{0.4992}$

From the above, we see that as the number of term "n" increases from 4 to 8, the sum of the term, S_n approaches 0.5.

Therefore, if the common ratio, r is a fraction such that $-1 < r < 1$, then the value of r^n approaches zero as "n" tends to ∞

Hence, S_n becomes

$$S_n = \frac{a\,(1 - r^n)}{1 - r}$$

$$S_n = \frac{a\,.}{1 - r},$$

means that the sum to infinity of G.P which is used to find the sum to infinity of any given G.P with a fractional ratio.

(e) The sum of the whole G.P is

$$S_n = \frac{a\,.}{1 - r}$$

$$= \frac{{}^1/_3\,.}{1 - {}^1/_3}$$

$$= {}^1/_3 \div {}^2/_3$$

$$= {}^1/_3 \times {}^2/_3 = {}^1/_2 = \mathbf{0.5}$$

Sum of the G.P is = 0.5 as n → ∞

REVISION EXERCISE 18.02:

1. Calculate the sum to infinity of the G.P $-^3/_4, -^3/_{40}, -^3/_{400}, -^3/_{4000} \cdots$

2. The sum to infinity of a G.P is 30. Calculate its common ratio if its first term is 48.

3. Find the sum of the first 9 terms of A.P:-24 ,-18 ,-12 . . . .

Chapter Nineteen

STATISTICS

19.01 **STATISTICS**

This is the collection of numerical data. This is also the aspect of decision making which has to do with numerical information. Decisions are made on the basis of available information supplied or implied. Some major decision makers include; the government, industries, administrative researchers etc.

Moreover, statistics is a scientific method of decision making under uncertainty when numerical data (*information in numeral form*) and calculated risks (because we are uncertain of what will actually happen are involved), though not every decision making requires the use of statistics.

POPULATION

This is a collection of the individual items (people or things) which are to be observed in a given problem or situation, for instance all the possible numbers of dots on the faces of a die in Ludo game thus (*1,2,3,4,5,6*). A population can be finite, infinite, countable or uncountable. Population is also related to sample, so in an infinite population, sample can be used

SAMPLE: This is a part of a population observed for the purpose of making scientific statement or taking decision about the population. Sample can be random, for instance, tossing of coin, throwing of dice, drawing slips (of papers) from a container or selection of members of the sample from a population. Sample can be purposive {if any of the above listed option is/are not applicable}

VARIABLE AND OBSERVATION

This is a characteristics possessed by the members of a population. It can be in the form of integers or real numbers. Example age, weight height etc the natural variable that its value is unpredictable is called random variable.

We have discrete variable {which takes only numbers} and its collection of such number is finite and countable example number of students, size of a family. Continuous variable takes any kind of real numbers. It is measured and not countable.

However, it can be observed in discrete forms example, weight, age, time etc in whole or fraction.

PARAMETER AND STATISTIC

A parameter is a characteristic of a population which helps to summarize information about the population with respect to variable under study for instance measure of location (*Central Tendency*) and Measure of Dispersion.

A statistic is to a sample as a parameter is to population.

FREQUENCY: This is the number of times each value or group of values occurs.

EXAMPLE 19.00: Price of single goat by four (4) buyers thus: N2000 N3000, N2000 and N4000.

However, the frequency of N2000 is 2 while N3000 is 1 and N4000 is 1.

Items	Frequency
N2000	2
N3000	1
N4000	1 .
	4 buyers.

PIE CHART

This is a circle divided into sectors. The circle represents the total data and each sector is proportional to its relative size.

EXAMPLE 19.01:

Ministry	Value	Degree
Commerce **(C)**	75.3	96.4^0
Transport **(T)**	90.4	115.7^0
Power **(P)**	115.5	147.8^0
	281.2	

Find the percentage for commerce:

$$\frac{75.3}{281.2} \times \frac{360^0}{1} = \mathbf{96.4^0}$$

For Transport

$$\frac{90.4}{281.2} \times \frac{360^0}{1} = \mathbf{115.7^0}$$

For Power:

$$\frac{115.2}{281.2} \times \frac{360^0}{1} = \mathbf{147.8^0}$$

∴   **C + T + P** = 359.96

= **360⁰**

Pictorically

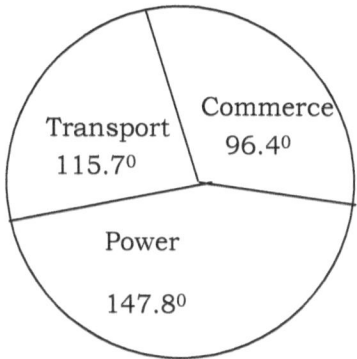

EXAMPLE 19.02:

With the below data find the percentages

Items	Value	Percentages	Degree
Maize **M**	102.4		
Groundnut, **G**	50.3		
Citrus, **C**	115.4		

SOLUTION

Since $102.4 + 50.3 + 115.4 = 268.1$

For **M**

$$x\% \text{ of } 268.1 = 102.4$$

$$\Rightarrow \frac{x}{100} \times \frac{268.1}{1} = 102.4$$

$$\frac{268.1}{100} x = \frac{102.4}{1}$$

Cross multiplying,

x = 38.2%

For **G**, $x\%$ of $268.1 = 50.3$

$$x = ?$$

$$\frac{x}{100} \times \frac{268.1}{1} = 50.3$$

$$\frac{268.1}{100} x = 50.3$$

x = 18.80%

For **C**

$$x\% \text{ of } 268.1 = 115.4, x =?$$

$$\frac{x}{100} \times \frac{268.1}{1} = 115.4$$

$$x = \frac{115.40}{268.1} = \mathbf{43.0\%}$$

∴ Totally the percentages and checking whether equal to 360°

$\Rightarrow$ 38.2 + 18.8 + 43.0 = **100.0%**

Representing with degrees

Maize will be $\quad \frac{102.4}{268.1} \times \frac{360}{1} = 137.5^0$

Groundnut will be $\quad \frac{50.3}{268.1} \times \frac{360}{1} = 68.5^0 \quad$ } 359.95 ≅ **360⁰**

Citrus will be $\quad \frac{115.4}{268.1} \times \frac{360}{1} = 154.95^0$

REVISION EXERCISE 19.00: The grades of thirty six (36) students in a Mathematics test are shown. How many students had excellent?

ANSWERS TO REVISION EXCERCISES

REVISION EXERCISES 1.00

SOLUTION

1. $(0.4)^2 = (^4/_{10})^2 = \{4\}^2/\{10\}^2 = 4 \times 4/ 10 \times 10 = {}^{16}/_{100}$

2. $(^{64}/_{27})^{-2/3} = {}^1/(^{64}/_{27})^{2/}{}_3$

But $\left(^{64}/_{27}\right)^{2/_3} = \left(\sqrt[3]{^{64}/_{27}}\right)^2$

$= \left(4/3\right)^2 = 4 \times 4 / 3 \times 3 = {}^{16}/_9$

$\therefore {}^1/\left(^{64}/_{27}\right)^{2/_3} = {}^1/{}_{16}/_9$

$= {}^9/_{16}$ (*the inverse of* $^{16}/_9$)

REVISION EXERCISES 1.01

SOLUTION

1. $\log_{7-1} 49 = x$

   $7^{-1(x)} = 49$

   $7^{-x} = 7^2$ (*Same base*)

   $-x = 2$

   $\therefore x = \mathbf{-2}$

2. $\log_5 25 = x$

   $5^x = 25$

   $5^x = 5^2$ (*Same base*)

   $\therefore \quad x = \mathbf{2}$

REVISION EXERCISES 1.02

SOLUTONS

1. 40.9 x 69.32

    $= 10^{1.6117} \times 10^{1.8408}$

    $= 10^{\{1.6117+1.8408\}}$

    $= 10^{3.4525}$

Finding the Antilog of .45 under 2 differencc 5. This is equal to 2834

Since the characteristics is 3,

then we move { 3+1 = 4 } places of decimals to the right,

2834 becomes **2834**.

REVISION EXERCISE 1.03:

1. $\dfrac{\{3.68\}^2 \times 6.705}{\sqrt{0.3581}}$ {*SSCE 1995*}

2. $\sqrt{\dfrac{0.897 \times 3.536}{0.00249}}$ { *SSCE 1994*}

SOLUTION

1.

No	log
$\{3.68\}^2$	$0.5638 \times 2 = 1.1316$
6.705	$0.8264 = 0.8264$
	$1.9580 = 1.9580$
from the denominator	
0.3581	$\overline{1}.5540 \div 2 = \overline{1}777$
	Having the characteristics as 2, then we go
	$\{2 - 1 = 1\}$ place of decimals from the left
	So , 1517. becomes **151.7**

SOLUTION

No	log
0.897	$\overline{1}.9528$
3.536	$\underline{0.5485}$ {+}
	0.5013
from the denominator	
0.00249	$\underline{\overline{3}.3962}$ { - }
	$\overline{3}.1051$ { ÷} 2

$$\sqrt{\dfrac{0.897 \times 3.536}{0.00249}}$$

Finding the Anti log of .55 under 2 difference 6 is 3569. Having 3 as the characteristics, then 3569 becomes { 3-1 = 2 } place of decimals = **35.69**

REVISION EXERCISE 2.00:

SOLUTION:

$\mu = 84$

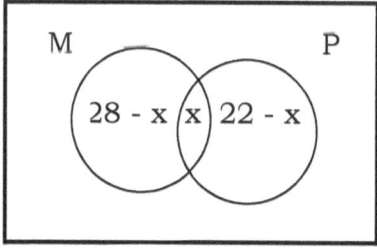

$$28 - x + x + 22 - x = 42$$

Bringing like term together

$$28 + 22 - x = 42$$

$$50 - x = 42$$

$$50 - 42 = x$$

$\therefore$ $x = 8$

This means, for Mathematics only

$28 - 8 = 20$ students will be present while for Physics

$$22 - x = 22 - 8$$

= 14 students will be present

(2

μ

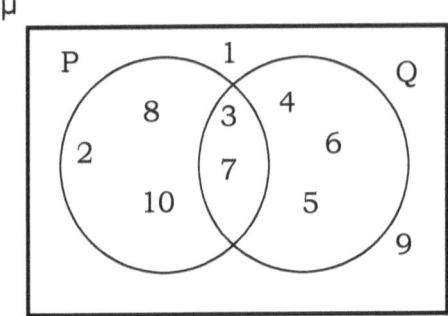

Having the above Venn diagram, it shows a Universal set , μ of integers and its subset P and Q.

The elements of the following

{i} PUQ

{ii} P∩Q

{iii}μ∩Q

{iv}PUμ

{v} P∩μ

However,

n(P) = { **8, 2, 10, 3, 7** }

n(Q) = { **4, 6, 5, 3, 7** }

n(μ) = {**1, 2, 3, 4, 5, 6, 7, 8, 9, 10** }

n(PUQ)ᶜ = { **1, 9** }

So, PUμ = { **2, 3, 4, 5, 6, 7, 8** }

P∩Q = { **3, 7** }

μ∩Q = { **3, 4, 5, 6, 7** }

PUμ = { **1, 2, 3, 4, 5, 6, 7, 8, 9, 10** } = μ

P∩μ = { **2, 3, 7, 8, 10** }

REVISION EXERCISE 3.00:

A machine was bought in 1990 for **₦5,500**. It was estimated to be **₦500** after 4 years of use. Then calculate the depreciation.

SOLUTIONS

Using the formula,

$$D_{rate} = \frac{\text{cost - Estimated /residual value}}{\text{years of useful life}}$$

= 5500 - 500/4

= 5000/4

= **1250**

REVISION EXERCICES 4.00:

1. The thickness of a book varies directly with the number of pages in the book. A book is 1.8cm thick and contains 350 pages. What is the thickness of the first 84 pages of the book?

2. If r is inversely proportional to h, and r =5 when h = 12.

   {a} find r when h = 20 { b} find h when r = 45.

3. x varies jointly as y and z when y=9 and z = 2

   {a} find the relation between x, y and z {b} find x when y = 14 and z= 12

1. The thickness of a book,**T** varies directly with the number of pages in the book, **N**. A book is 1.8cm thick and contains 350 pages. What is the thickness of the first 84 pages of the book?

2. If r is inversely proportional to h, and r =5 when h = 12.

   {a} find r when h = 20 { b} find h when r = 45.

3. x varies jointly as y and z when y=9 and z = 2

   { a}find the relation between x, y and z {b} find x when y = 14 and z= 12

SOLUTION

1. $T \alpha N$

   $T = kN$, where k = the constant of proportionality

   $k = T/N$, k = 1.8/350 = 0.005142

Finding **T** when **N** = 84 pages and having k as 0.005142, then, $T = kN$

   $T = 0.005142 \times 84 = \mathbf{0.432}$

2. r α 1/h , r = k/h.

So,   k = hr

   i.e 12 x 5 = 60. { a} r = k/h, $\mathbf{r}$ = 60/20 = **3**.

   {b} hr = k , $\mathbf{h}$ = k/r = 60/45 = **1.33**

3. x= y/z

   x= ky/z

   ky = zx

   k = zx/y

   $= 2 \times 27/_9 = 6$

{a} $\mathbf{x = 6y/z}$

{b} k = zx/y

So,   x = ky/z

   = 6x14/12

$x = 84/12 = 7$

REVISION EXERCISE 5.00:

SOLUTIONS

1.

i} Given $u^2+18u+72$,

We find the factors of 72 such that addition or subtraction of the factors gives 18 i.e 6 and 12.

$d/4 \{ x2+6x \}+\{ 12x+72 \}=0$

$x\{ x+6\} +12\{ x+6\} = 0$

$\{x+12\}\{ x+6\}.$

ii} Given $12x^2+14x-20$,

$\{12x^2 + 24x\} \{-10 - 20 \}$

$6x\{ 2x +4 \}-5\{ 2x + 4 \}$,the common factor among the two terms are

$\{ 2x + 4 \}$

$d/4 \{ 6x - 5 \}\{ 2x +4 \}$ are the product.

iii} Given $16x^2 \text{-} 9$

$4x\{4x-3 \}+3\{ 4x-3\}$

$= 16x^2 \text{-} 12 +12x-9$

$= 16x^2 - 9$, So the products are { **4x + 3** }{ **4x + 3** } since the common factors are { 4x-3}.

iv} Given $25y^2 - 4$

$5y\{ 5y - 2 \} + 2\{ 5y - 2 \}$

$= 25y^2\text{-}10y + 10y\text{-}4$

$= 25y^2\text{-}4$, so the products are { **5y + 2**}{**5y-2** } since the common factors are { 5y-2}.

Using AC method, solve $2x + 7x - 15$

We find a $=2, b=7$ and c $=-15$

ac= $2\{-15\} = -30$, b= 7

thus, mn $= -30$ and m + n = 7

We calculate integer pairs as;

mn	m+n
1{ -30} = -30	1+{ -30} = -29
2{ -15} = -30	2+{ -15} = -13
3{ -10} = -30	3+{ -10} = -7
5{ -6} = -30	5+{ -6} = -1
6{ -5} = -30	6+{ -5} = 1
10{ -3}	

$\qquad$ =-30 $\qquad\qquad$ 10+{-3} = 7

There is no need to go any further; we see that 10 and -3 have a sum of 7.

So, **m=10 and n =-3**

Therefore, $2x^2-7x-15$ is factorable.

REVISION EXERCISE 6.00:

SOLUTIONS:

1. $x-3x = 7$

   $-2x = 7$

   x =-7/2

2. $6x-x + 2 = 0$

   $5x + 2 = 0$

Bringing like terms together,

   $5x =-2$

   x =-2/5

3. $x-4x + 4 =-5$

   $-3x + 4 =-5$

Bringing like terms together,

   $-3x =-5-4$

   $-3x =-9$

Canceling out the minus signs,

$3x = 9$

x = 3

4. 2-3x + 1= 0

Bringing like terms together,

$-3x = -1-2$

$-3x = -3$

Canceling out the minus signs,

$3x = 3$

x = 1

REVISION EXERCISE 7.00:

SOLUTION

1. $3x + 5y = 4$... equation i
    $4x + 3y = 5$... equation ii
    $12x + 20y = 16$..equation iii
    $12x + 9y = 15$..equation iv
    $11y = 1$
    y = $^{1/11}$

To get x, substitute the value of y into equation ii,

$4x + 3y = 5$ becomes

$4x + 3 \{ ^{1/11}\} = 5$

$4x = {}^{5/1 - 3/1}$

$4x = 52/11 \div 4/1$

$x = {}^{13/11}$ { *JAMB 1982* }

2. $x + y = {}^{3/2}$

$x - y = {}^{5/2}$

The above equation can be put in the form,

2x+2y=3... equation i
2x-2y= 5 .... equation ii
2x+4y=6
2x-4y = 10
8y=-16
y =-2

To get the x, we substitute the value of y into equation ii,

2x-2y = 5

2x-2 {-2 } = 5

2x-4 = 5

2x = 9

x = $^{9/2}$ = **4**$^{1/2}$

3. x + y = 2
3x-2y = 1
3x-3y = 1
3x-2y = 1

-5y = 5
y = 5/-5

y =-1

To get x , we substitute the value of y into equation ii,

$3x-2y = 1$

$3x-2 \{-1\} = 1$

$3x-2 = 1$

$3x + 2 = 1$

$3x = 1-2$

$3x =-1$

x =-1/3   { **SSCE 2007**}

4. $3x-2y=21$

   $4y+5x=5$

Rewriting the above equation so as to align x and y ,

$$3x-2y = 21$$
$$5x+4y = 5$$
$$\overline{15x-10y = 105}$$
$$15x+12y =15$$
$$\overline{-22y = 90}$$

y = 90/-22

To find x, we substitute the value of y into equation ii,

$15x + 4y = 5$

$15x + 4\{ 90/-22 \} = 5$

$330x + 360 = 110$

330x = 110-360

330x =-250

x =-250/330   **{ SSCE 1999 }**

REVISION EXERCISES 8.00:

1) Think of a number, subtract 12 from its square the result is 30 added to number, and find the number.

SOLUTION

Let the number be = x

$x^2 - 12 = 30 + x$

$x^2 - x - 12 - 30 = 0$

$x^2 - x - 42 = 0$ forms a quadratic equation where a = 1, b =-1, c =-42

Using formula method

$$x = \frac{-b \pm \sqrt{b^2 - 4ac}}{2a}$$

$$= \frac{-1 \pm \sqrt{(-1)^2 - 4(1)(-42)}}{2(1)}$$

$$= 1 \pm \sqrt{1-168}/2$$

$$= 1 \pm \sqrt{169}/2$$

$$= \frac{1 \pm 13}{2}$$

$x_1 = \dfrac{1+13}{2} = 7$

$x_2 \dfrac{1-13}{2} = -6$

Verifying

$7^2 = 49$

$\dfrac{-12}{37}$ which is $30 + 7$

OR

$(-6)^2 = 36$

$\dfrac{-12}{24}$ which is $30 +-6$

2] Seven years ago the age of a father was three times that of the son , but in six years time the age of the son will be half the father. Representing the present age of the father and don by x and y respectively ,the two equations relating x and y are:

SOLUTION

Let the fathers age be x

Let the sons age be y

In 7years ago

$x-7$ $= 3 \{ y - 7 \}$

$= 3y-21$

$$x-y = 3y-21$$

Bringing like terms together,

$$x-3y = -21 + 7$$

$$x-3y = -14$$

In 6years time ,

$$\tfrac{1}{2}\{x + 6\} = \{y + 6]$$

$$x + 6 = 2\{y + 6\}$$

Multiplying out,

$$x + 6 = 2y + 12$$

Bringing like terms together,

$$x-2y = 6, \quad \text{the two equations are}$$

$$x-3y = -14$$

$$x-2y = 6 \qquad \{ \textit{JAMB 1982}\}$$

REVISION EXERCISE 9.00:

1) Given $\dfrac{P}{\sqrt{2}} = \sqrt{\dfrac{r\,2}{r + q}}$ make r the subject

SOLUTION

$$\frac{P}{\sqrt{2}} = \sqrt{\frac{r\,2}{r+q}}$$

$\Rightarrow \quad \left(\dfrac{P}{2}\right)^2 = \dfrac{r\,2}{r+q}$

$P^2 (r+q) = r.2$

$P^2r + P^2q = 2r$

Bringing like terms together

$P^2q = 2r - P^2r$

$P^2q = r(2 - P^2)$

$\therefore \quad \mathbf{r = \dfrac{P^2q}{2 - P^2}}$

2) Make x the subject given

$$\dfrac{1 - ax}{1 - ax} = \dfrac{P}{q}$$

SOLUTION

Cross multiplying

$$\dfrac{1-ax}{1-ax} \overset{=}{\times} \dfrac{P}{q}$$

$q(1-ax) = P(1-ax)$ equation (i)

$q - qax = P - Pax$ equation (ii)

Bringing like terms together

$-qax + Pax = P - q$ equation (iii)

$$-x(qa - Pa) = P - q \qquad\qquad \text{equation(iv)}$$

Multiply through by-1

$$-1(-x)(qa-Pa) = -1(P - q)$$

$$x(qa - Pa) = P + q$$

$$\therefore \quad x = \frac{P + q}{qa - Pa}$$

$$x = \frac{P + q}{a(q - P)}$$

OR

From

$$P(1-ax) = q(1-ax)$$

$$P - Pax = q - qax$$

$$-Pax + qax = q - P$$

$$-x(aP - qa) = q - P$$

$$-x = \frac{q - P}{a(P - q)}$$

$$\therefore x = \frac{(q + P)}{a(P - q)}$$

3) Make c the subject, if

$$r = 1 - \frac{a}{5}\left[\frac{b-3c}{7}\right]$$

SOLUTION

Solving for the bracket first,

$$\frac{b}{1} - \frac{3c}{7}$$

$$\Rightarrow \quad \frac{7b - 3c}{7}$$

$$\therefore \quad r = 1 - a/5\left(\frac{7b-3c}{7}\right)$$

$$r = \frac{1}{1} - \frac{7ab + 3ac}{35}$$

$$r = \frac{35 - 7ab + 3ac}{35}$$

Cross multiplying

$$35r = 35 - 7ab + 3ac$$

$$35r - 35 + 7ab = 3ac$$

Factorizing

$$35(r-1) + 7ab = 3ac$$

$$3ac = 35(r-1) + 7ab$$

$$\therefore \quad c = \frac{1}{3a}\left(35(r-1) + 7ab\right)$$

$$= 1/3\left(\frac{35(r-1) + 7b}{a}\right)$$

REVISION EXERCISE 10.00:

2) Solve for k

$$2k-2 \leq \frac{k+5}{2}$$

SOLUTION:

$$\frac{2k-2}{3} \leq \frac{k+5}{2}$$

Cross multiplying

$$2(2k - 2) \leq 1(k + 5)$$

$$4x-4 \leq k + 5$$

Bringing like terms together

$$4k - k \leq 5 + 4$$

$$3k \leq 9$$

$$\therefore \quad k \leq 9/3$$

k ≤ 3

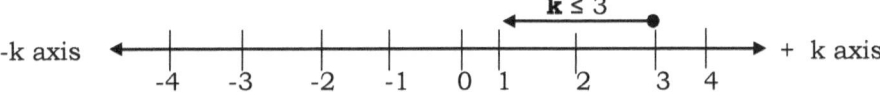

REVISION EXERCISE 11.00:

CONGRUENCY

1.

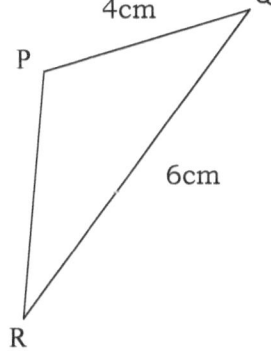

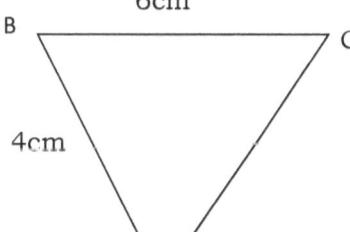

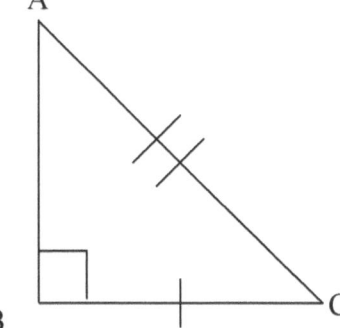

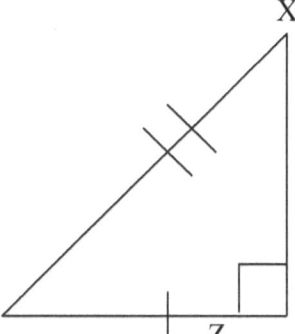

1. In triangle PQR and CBA,

PQ = BA

= 4cm

QR = BC

= 6cm

Angle PQR = Angle CBA

= **60⁰**

2. In the Right angled triangles, ABC and XZY

AC = XY

and

BC = ZY

Angle ABC = Angle XZY

= **90⁰**

So, triangle ABC = triangle XZY { R H S }

REVISION EXERCISE 12.00:

SOLUTION

1. Sum of exterior angle and the interior angles of a polygon = 180⁰

Let the size of the exterior angle be x,

then, $x^0 + 108^0 = 180^0$

So, $x^0 = \{180\text{-}108\}^0$

= **72^0**

2. Size of each exterior angle = $360^0/n$

    where n = number of sides of the polygon

    So ,each angle of the pentagon will be , $360/5 =$ **72^0**

3. Size of each exterior angle , = $360^0/n$

    So, each angle of the 8-sided { octagon} will be, $360^0/8$ = **45^0**

4. **<u>For the 10 sides,</u>**

Using the formula, $S_{int} = [2n\text{-}4\,]$rt angle

$= [\,\{\,2x10\} - 4\,]x\ 90^0$

$= [\,20 - 4\,]\ x\ 90^0$

$= 16\ x\ 90^0$

= **1440^0**

For the 11 sides,

We still use the above formular,thus

$S_{int}\quad = [\,2n - 4\,]$rt angle.

$= [\,\{\,2\ x\ 11\,\} - 4\,]x\ 90^0$

$= [\,22 - 4\,]\ x\ 90^0$

= 18x 90⁰

= **1620⁰**

REVISION EXERCISE 13.00

SOLUTION

{1} {a} Given the diagram as :

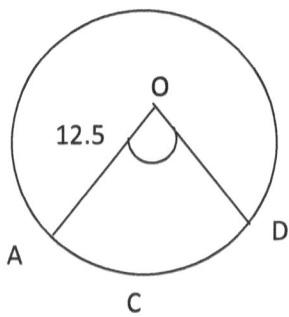

AD   = d Sin θ/2

= 25 x Sin 32⁰

= 25 x 0.529 = 13.2cm

{ b}   Perimeter of the segment ACD = AD + Arc ACD

= 13.2 + 640/360 X 3.142 X 25

= 13.2+ { 0.177x 3.142x 25 }

= 13.2+ 13.2 + { 13.96}

= **27.16 cm**

{2}

BMC = length of the chord

So using Pythagoras theorem, we can find the first half, thus

$$\overline{\{MC\}}^2 \quad = \overline{\{AC\}}^2 - \overline{\{AM\}}^2$$

$$= \{9.5\}^2 - \{5\}^2$$

$$= 90.25 - 25$$

$$= \sqrt{65.25}$$

$$= 8.1cm$$

So, $\overline{BMC}$ = 2 x 8.1 = **16.2cm**

SOLUTION

1. Sum of exterior angle and the interior angles of a polygon = 180^0

   Let the size of the exterior angle be x,

   then, $x^0 + 108^0 = 180^0$

   So, $x^0 = \{180-108\}^0$

   = **72^0**

{2} Size of each exterior angle = $360^0 / n$

   where n = number of sides of the polygon

So ,each angle of the pentagon will be , 360/5 = **72⁰**

3. Size of each exterior angle , = $360^0/n$

So, each angle of the 8-sided { octagon} will be, $360^0/$ 8 = **45⁰**

4. <u>For the 10 sides,</u>

Using the formula, S_{int} = [2n-4]rt angle

= [{ 2x10} - 4]x 90^0

= [20 - 4] x 90^0

= 16 x 90^0

= **1440⁰**

For the 11 sides,

We still use the above formular,thus

S_{int} = [2n - 4]rt angle.

= [{ 2 x 11 } - 4]x 90^0

= [22 - 4] x 90^0

= 18x 90^0

= **1620⁰**

REVISION EXERCISE 14.00:

REVISION EXERCICES 15.00:

Given a triangle, find the area by Hero's formula

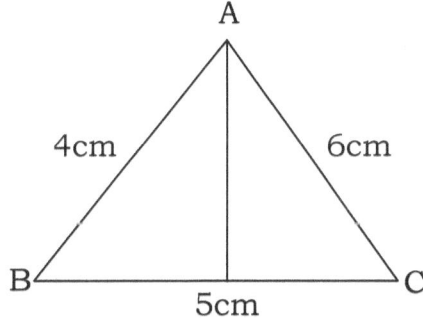

By Hero's formula, the area is solved as : s = a + b + c/ $_2$

$= 5 + 6 + 4 /_2 = 15/_2 = 7.5$

A = s { s - a } { s - b} { s - c }

$= \sqrt{ 7.5 \{ 7.5\text{-}5 \} \{ 7.5\text{-}6 \} \{ 7.5\text{-}4 \} }$

$= \sqrt{ 98.4375 }$

= **9.92cm²**

2. Using Hero's formula, solve:

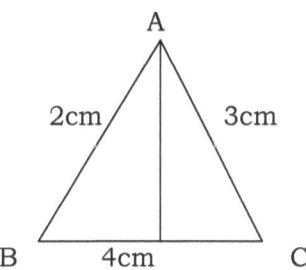

S $\quad = a + b + c /_2$

$\quad\quad = 4+3+2 /_2 = 9/2 = 4.5$

$A = \sqrt{s\{s-a\}\{s-b\}\{s-c\}}$

$A = \sqrt{4.5\{4.5-4\}\{4.5-3\}\{4.5-2\}}$

$= \sqrt{4.5\{0.5\}\{1.5\}\{2.5\}} = \sqrt{8.4375} = \mathbf{2.89cm^2}$

3. Given a triangle below, find the area.

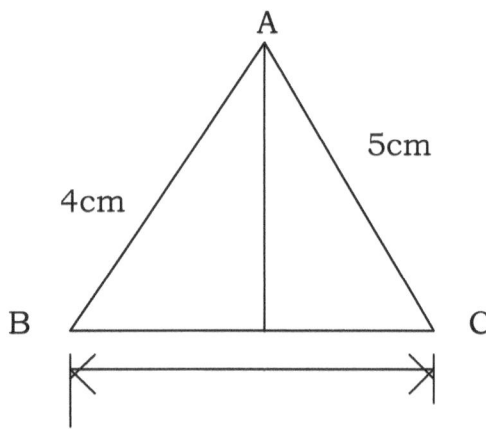

Area = ½ { b } x h = ½ { 10 } x 3 = { 5 x 3 } = **15cm²**

REVISION EXERCISE 17.00

SOLUTION

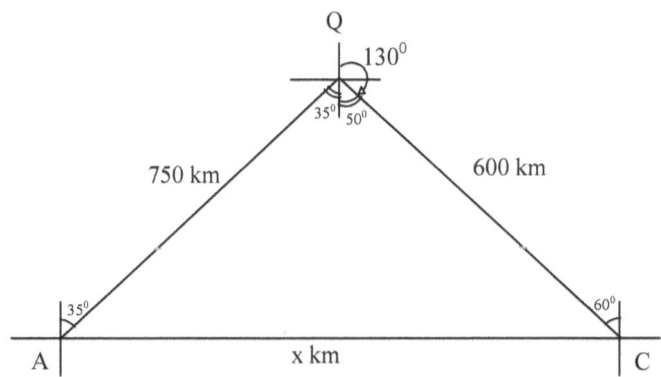

If 600 km = 1hr

x km = 1.25hrs = (1¼hrs)

Cross multiplying

x km = **750**

Also, if 400 km = 1hr

Then x km = 1.5hrs = (1½ hrs)

So, x km = **600**

The above, diagram can be redrawn as

i. Using Cosine Rule,

$$b^2 = a^2 + c^2 - 2ac \cos B^0$$

$$= (600)^2 + (750)^2 - 2 \times 600 \times 750 \times \cos 85^0$$

$= 360{,}000 + 562{,}500 - 900{,}000 \times 0.08715$

$= 922500 - 78440.168$

$= 844059.8$

$b = \sqrt{844059.832}$

$= 918.7$

= **919 km**

Using the above redrawn triangle, we can use the Sine Rule, thus;

$$\frac{c^0}{\text{Sin C}} = \frac{b^0}{\text{Sin B}}$$

$$\frac{750^0}{\text{Sin C}} = \frac{919^0}{\text{Sin 85}}$$

Cross multiply

$919 \, \text{Sin C}^0 = 750 \times 0.99619$

$\text{Sin C}^0 = 747.146 \div 919$

$= 0.81299$

$C^0 = \text{Sin}^{-1} 0.81299$

$= 54.3899$

= **54^0**

So, the bearing of C from A is $35^0 + 41^0 =$ **076^0**

2. X, Y and Z are three points on the horizontal ground such that /xy/ = 5m, /yz/ = 8m the bearing of X from Y is 030^0 and the bearing of Z from Y is 140^0

(a) find correct to one decimal place the;

{ i.}   distance between X and Z

{ ii}   Bearing of X from Z

(b) How far east of Y is Z ?

(*correct to the nearest meter*)

SOLUTION

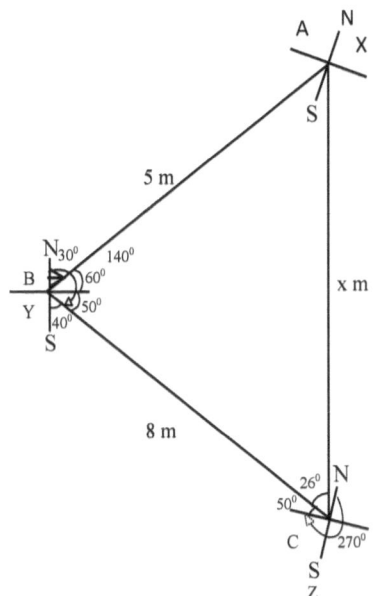

$$60^0 + 50^0 = 110^0$$

So, $B^0 = 110^0$

b^2 = a² + c² - 2ac Cos B⁰ (*By Cosine Rule*)

= 8² + 5² - 2 x 8 x 5 Cos 110⁰

$$= 64 + 25 - 80 \, (-0.3420)$$

$$= 89 - 27.36$$

$$b^2 = \sqrt{116.36}$$

$$= 116.36$$

$$= 10.78m$$

$$= \mathbf{10.8m}$$

However

$$\frac{5}{\text{Sin } C^o \, (Z)} = \frac{8}{\text{Sin } A^o \, (X)} = \frac{XZ \, (X)}{\text{Sin } B^o \, (Y)}$$

$$\frac{5}{\text{Sin } C^o \, (Z)} = \frac{10.78}{\text{Sin } 110^o} \qquad \text{equation (i)}$$

OR

$$\frac{5}{\text{Sin } A^o \, (X)} = \frac{10.78}{\text{Sin } 110^o} \qquad \text{equation (ii)}$$

$C = Z = 25.8^o$ from the equation (i)

Verifying

$$10.78 \, \text{Sin } C^o = 5 \times \text{Sin } 110^o$$

$$10.78 \, \text{Sin } C^o = 5 \times 0.9396$$

$\text{Sin } C^o = 0.4358$

So, $C^o = \text{Sin}^{-1} 0.4358$

$$= 25.83$$

$$= \mathbf{26^o}$$

The bearing of X from Z is

$= 50^0 + 25.8^0 + 270^0$

$=245.80$

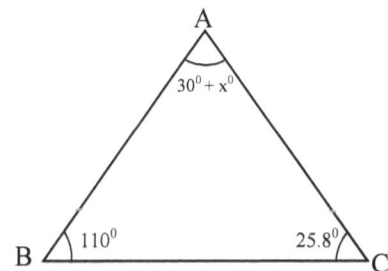

So, $25.8^0 + (30 + x)^0 + 110^0 = 180^0$ (*Sum of angle in a* $\triangle$)

$165.8^0 + x^0 = 180^0$

$x^0 = 180^0 - 165.8$

$= 14.2^0$

$260^0 - 14.2^0$

$= 345.8^0$

This is the bearing of X from Z.

SOLUTION

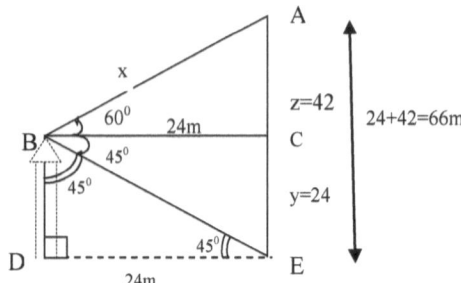

(i) 24 + 42 = 66m

     = Height of the mast.

Tan 45^0 = $\dfrac{\text{Opp.}}{\text{Hyp.}}$ (**SOHCAHTOA**)

Tan 45^0 = $\dfrac{24}{y}$

Cross multiply.

y = $\dfrac{24}{\text{Tan } 45^0}$

     = 24/1

∴ **y = 24m**

Tan 60^0 = $\dfrac{z}{24}$

z   = 24 x Tan 60^0

     = 24 x 1.7320

     = 41.56

z = 42m

Cos 60° = $\frac{24}{x}$ (**SOHCAHTOA**)

x Cos 60° = 24

x = $\frac{24}{Cos\ 60°}$

x = $\frac{24}{0.5}$

x = 48m

REVISION EXERCISE 18.01:

1. Given l =-88, n = 16, s =-448, find "a" and "d"

2. How many terms of the series 42, 39, 36 . . . must be taken that the sum may be 312?

3. Given a series 1, $^4/_3$, $^5/_3$ as an A.P. find:

   (a) 10th term

   (b) Sum up to the 16th term { **NECO 2009**}

SOLUTION 18.01:

{1} Having the formula as ;

$\qquad$ l = a +{ n-1} d

$-88 = a + \{ 16\text{-}1 \} d$

$-88 = ,a +15d$ equation 1

$15d = -88 - a$

$d = d\text{-}88 - a \ /15$

$N_{16th} = a +\{ n\text{-}1\} d$

$S = {}^{n/2}\{ 2a + [n - 1]d \}$

$-448 = {}^{16/2} \{ 2a + 15d\}$

$-896 = 16\{ 2a+15d\}$

$-896 = 32+ 240d$ equation 2

meanining that equation {1} and {2} turns to be a simultaneous equation.

$a +15d = -88$ equation 1

$32a + 240d = -896$ equation 2

$32a + 480d = -2816$ equation 3

$32a + 240 d = -896$ equation 4

$240d = -1920$

$d = -8$

To get " a ",we substitute value of " d " into equation 2

$32a + 240d = -896$

$32a +249\{-8\} = -896$

32a +-1920 =-896

32a =-896+1920

32a = 1024

So, a = 32/1024

   = **32.**

{2} Using , S_n = n { a + 1 } /2

   312 = n { 42+ 36} /2

Cross multiplying,

   624 = 78n

So,   **n = 8**

REVISION EXERCISE 18.02:

SOLUTION:

1. Using, b/a= b/c

   -3/40 x 4/-3 =-3/400 x40/-3

   -10 =-10

So, common ratio is 10.

Sum to infinity of the G.P is by formula,

   S = a/1-r

DAVIDSON C. OKOKO & CHIDOZIE C. OKOKO

$$=\frac{-3/4}{1-10}$$

$$=\frac{-3/4}{-9}$$

S = 1/12

2. Using the formula,

$$S = a / 1 - r$$

$$30 = 48 / 1 - r$$

cross multiplyimg ,

$$30\{ 1-r\} = 48$$

$$30 - 30r = 48$$

Bringing like terms together,

$$-30r = 48 - 30$$

$$-30r = 18$$

r = 18/-30

Is   b-a = c - b ?

$$4/3 - 1/1 = 5/3 - 4/3$$

$$1/3 = 1/3$$

So,   d = 1/3

$$10_{th} \text{ term} = a + \{ n-1 \} d$$

= 1 + 9d

= 1+ 9 { 1/3}

= 1+ 3

= **4**

Using the formula,

$S= \ _{n/2} \{ 2a+ [n-1] d\}$

$S_{16th} = \ _{16/2} \left[2\{1\} +[16-1]1/3 \right]$

= 8{ 2 +15x1/3}

= 8{ 2 + 5}

= 8 x 7

= **56**

Is   b-a = c-b ?

-18-24 =-12-18

-18+ 24 =-12 + 18

6 = 6

S_9 = n/2{ 2a+[n-1]d}

= $^{9/2}$ { 2[-24]+[9-1]x 6 }

= 4.5 {-48 + 8 x 6}

= 4.5 {-48 + 48}

$= 4.5 \times \{ 0 \} = 0$

$S_9 \quad = 0 \{ \textbf{\textit{NECO 2011}} \}$

REVISION EXERCISE 19.00:

The grades of thirty six (36) students in a Mathematics test are shown. How many students had excellent?

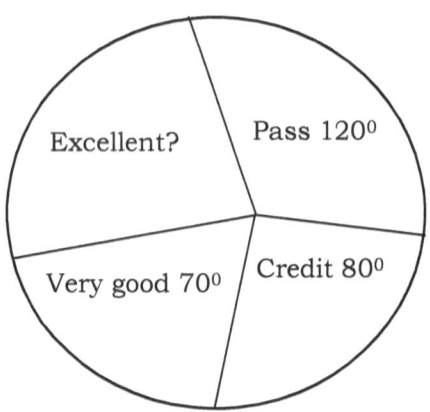

SOLUTION

Total number of students = 36

$\therefore \dfrac{\text{Number of students}}{\text{Total number of students}} \quad \times \dfrac{360}{1}$

$= \dfrac{x}{36} \times \dfrac{360^0}{1} = 120^0$

$\dfrac{360^0}{36} x = 120$

Cross multiplying

$x = 12$ students

$$\frac{x}{36} \times \frac{360}{1} = 80^0$$

$$\frac{360^0 x}{36} = \frac{80^0}{1} \qquad \{Cross\ multiplying\}$$

$x = 8$ Students

∴ We still have 16 students

So for Very Good

$$\Rightarrow \quad \frac{x}{36} \times \frac{360^0}{1} = 70^0$$

$$\frac{360^0 x}{36} = 70^0$$

Cross multiplying

$10x = 70$

$x = 7$ students

So, $12 + 8 + 7 = 27$ students

∴ 36(*total number of students*) - 27 = **9** students is the number of students representing Excellent.

2. Representing this in a Pie chart,

        25,000,000   = Agric, **A**
        20,000,000   = Education, **E**
        35,000,000   = Welfare, **W**

$$\frac{20,000,000}{100,000,000} = \text{Commerce, } \mathbf{C}$$

∴ For **A**,

$$\frac{25,000,000}{100,000,000} \quad \text{x} \quad \frac{360^0}{1}$$

$$= \frac{25}{100} \quad \text{x} \quad \frac{360^0}{1}$$

$$= 25 \text{ x } 36$$

$$= \mathbf{90^0}$$

For **E**,

$$\frac{20,000,000}{100,000,000} \quad \text{x} \quad \frac{360^0}{1}$$

$$= \frac{20}{100} \quad \text{x} \quad \frac{360}{1} = \mathbf{72^0}$$

For **W**,

$$\frac{35,000,000}{100,000,000} \quad \text{x} \quad \frac{360^0}{1}$$

$$= \frac{35}{100} \quad \text{x} \quad \frac{360^0}{1} = \mathbf{126^0}$$

For **C**,

$$\frac{20,000,000}{100,000,000} \quad \text{x} \quad \frac{360^0}{1}$$

$$= \frac{20}{100} \text{ x } \frac{360^0}{1} = \mathbf{72\%}$$

∴ **A** + **E** + **W** + **C** = 360^0 (*Sum of angle in a circle*)

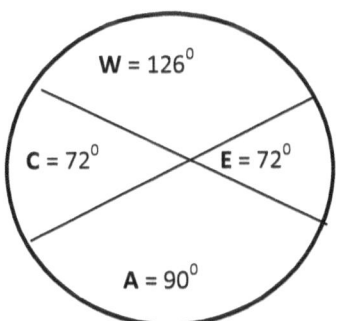

Bibliography

Adamu S.O et al (1997) <u>Statistics for Beginners Book I</u> SAAL Publication, Bodiga University of Ibadan{ P.15 - 352}

Adekunle O.Eyitayo et al (2006) <u>Computer Studies for Beginners I</u> Bounty Press Ltd Ibadan, Nigeria { P.27 }

Adu, D.B (2004) <u>Comprehensive Mathematics for Senior Secondary Schools</u> A Johnson Publishers Ltd,Surelere, Lagos {P.40}

Ale, S.O (2006) <u>Basic Concepts on Difficult Areas in Sec. Sch. Maths & Solutions to WASSCE, NECO & SSCE Registration</u> National Mathematical Centre Kaduna - Lokoja Rd Studu Kwali Abuja, Nigeria. Math Improvement Project (MIP) { P. 33}

Ajayi, O. W (2001) D.J <u>Revision keys on Mathematics for Senior Sec, Schools</u>, Daniel Jackson Publishers, 85 Lawson Rd, Surulere Lagos { P.34}

Borowski et al{ 2002},<u>Collins Dictionary of Mathematics.</u> Second Edition, Harper Collins Publishing London { P.277}

Donard Hutchison et al { 2005},<u>Beginning Algebra</u> 6th Edition Published by McGraw Hill, P. {388 , 390}

Hall. H.S (1951) <u>A School Algebra:</u> Macmillian and Co Ltd St Martins Street London P.307.

Igbokwe, D.I. et al (2005) <u>Science Teachers Association of Nigeria (STAN) Mathematics</u>, University Press Plc (Ibadan) Nigeria{P.62}

Jaggi V.P (2006) <u>Dictionary of Mathematics</u>, Educational Printing and Publishing Academies. Star Offset Printers, New Delhi India { P.7 - 305}

Macrae M.F et al{ 2005}, <u>New General Mathematics for West Africa,</u> Junior Secondary School 3 Third Edition University Press Limited Ibadan,Oyo State Nigeria{ P.172-175,135}

Okoko, D.C (2007), <u>Teach Yourself Investing</u> Published by AAA Capital Market & Real Estate Corporation, New York,USA {P. 27, 34)

Ugenyi, Chris (2006) <u>Master Job Aptitude Text, GMAT</u>. IEC Publication Bureau, #653 Ikorodu Road Opposite Mobil Filling Station, Mile 1 Lagos { P1,15 }

www.investopedia.com

www.icoachmath.com

http://mathforum.org.dr.math/.com

www.seriesmathstudy.com

en.wikipedia.org